LINEAR PROGRAMMING
for DECISION MAKING
AN APPLICATIONS APPROACH

DAVID R. ANDERSON
University of Cincinnati

DENNIS J. SWEENEY
University of Cincinnati

THOMAS A. WILLIAMS
University of Cincinnati

WEST PUBLISHING CO.
St. Paul • New York
Los Angeles • San Francisco

Library of Congress Cataloging in Publication Data

Anderson, David, 1941–
    Linear programming for decision making.

    Bibliography: p.
    1. Linear programming. 2. Decision-making.
I. Sweeney, Dennis, joint author. II. Williams, Thomas 1944–   joint
author. III. Title.
T57.74.A46   658.4'03   74–1138
ISBN 0–8299–0008–X

    4th Reprint—1978

TO LYNN, CHERRI, AND ROBBIE

*

PREFACE

The purpose of this book is to provide students, primarily in the fields of administration or economics, with an introduction to linear programming and its applications. The book is applications oriented. Theoretical topics do appear in the text where they are necessary in order to develop basic concepts, but each new theoretical development is followed immediately by an illustration of how it can be applied. Unique features of this book include one chapter devoted to typical problems in the functional areas of business, and another chapter devoted exclusively to implementation procedures for large-scale linear programming models.

Our intention is to provide the reader with a basic understanding of linear programming methodology and an honest appreciation of linear programming applications. The serious student will gain an understanding of linear programming that will enable him to work alone or with others in formulation, implementing, and interpreting linear programming models of real-world problems. For the student intending to pursue more advanced study, this book will provide a sound introduction to linear programming.

The mathematical prerequisite for a study of this book is a course in college algebra. In instances where we have found it desirable to use more advanced mathematical concepts, such as matrix algebra, the necessary material has been incorporated into the text where needed. By adopting this approach, we build the necessary mathematical skills of the reader as the linear programming topics are developed. Thus we avoid the usual practice of devoting a specific chapter and/or appendix to mathematical background before the student can see how this material relates to linear programming.

Most of the fundamental topics in linear programming are treated in the first four chapters of this book. Continuity is provided by introducing in Chapter 1 a linear programming

v

problem faced by Par, Inc., a small company engaged in the manufacture of golf bags. As new linear programming concepts are introduced in Chapters 2–4, they are immediately illustrated by showing how they can be applied to the Par, Inc. problem. Our experience in building a linear programming course around this material at the University of Cincinnati has been that students grasp new concepts much more quickly if the concepts can be related to a problem with which they are familiar. In addition, the Par, Inc. problem, which is a simplified prototype of a real-world problem, motivates student interest and is an aid to their intuition.

The applications orientation of the book becomes apparent in Chapter 1 as the Par, Inc. problem is introduced as a scenario of a problem faced by many manufacturing firms. In this initial chapter the reader is taken through the steps necessary to transform the verbal description of a problem into the formulation of the linear programming objective function and constraints. To illustrate the material, the Par, Inc. problem is formulated and solved using the graphical approach.

A computer code which is suitable for solving small to moderate sized linear programs is included in the appendix. The use of this linear programming code is illustrated by showing how it can be used to solve the Par, Inc. problem. Many instructors will wish to introduce the use of the computer for solving linear programming problems early in the course by assigning the appendix material immediately following Chapter 1. This is the sequence we have found to be most desirable. Other instructors, either because of lack of computer facilities or differing course objectives, may wish to delay or omit discussion of computer solution procedures. In such courses, the instructor can proceed directly from Chapter 1 to Chapter 2 with no loss of continuity.

The Simplex method is introduced in Chapter 2 by showing how it is used to solve the Par, Inc. problem formulated in Chapter 1. For pedagogical reasons of not wanting to digress from the main development, we consider only maximization problems with less-than-or-equal-to constraints in the first two chapters. As a result, the formulation exercises at the end of the chapters are limited to this type of problem.

In Chapter 3, the minimization problem, other constraint forms, and special topics such as infeasibility, unboundedness, alternate optima, and degeneracy are included. The important concepts of duality and sensitivity analysis are introduced in Chapter 4. While these are usually considered more advanced linear programming topics, they are very important in applications, and our presentation which relates these concepts to the Chapter 1 problem should enable the novice linear programming student to grasp the material.

Chapters 5 and 6, we believe, are a particular strength of the book and are devoted to more detailed applications of the linear programming skills and knowledge developed in the first four chapters. In Chapter 5 we have presented applications from each of the functional areas of business. Hence, most business students will find an application in their particular area of interest. Also, several of the classical linear programming applications such as the transportation, assignment, and diet problems are presented in this chapter. The applications in this chapter are all independent of one another. Thus the instructor may find it desirable to introduce some of these applications earlier in the course as a means of motivating the more theoretical material.

In Chapter 6, a production scheduling problem provides the basis for studying large-scale linear programming models. This chapter is a unique feature of the book and is designed specifically to acquaint students with the real problems associated with implementing a linear programming model. While the notation in this chapter is somewhat complex, we believe studying this chapter is an absolute necessity if one wants to develop an understanding of the data handling, computer solution and report generation steps of the implementation process.

Selected problems and a glossary are presented at the end of each chapter for the reader's use in reviewing the concepts presented in the chapter.

The text has been kept relatively short so that the material can be covered in a one-semester or accelerated one quarter course. By limiting study to Chapters 1, 2, 3, and portions of 5, the text can also be used for part of an introductory operations research or quantitative methods course that devotes only four or five weeks to the study of linear programming.

We would like to express our appreciation to Dr. John McClain, Dr. Carl Hamilton and Dr. Kenneth Rich for their suggestions during the development of this manuscript. We would also like to express our appreciation to our colleagues, students, and families who, in so many ways, have contributed to the successful completion of this project. We are especially indebted to Miss Jo Anne Hargis for the long hours she devoted to typing and retyping the drafts of this book.

DAVID R. ANDERSON
DENNIS J. SWEENEY
THOMAS A. WILLIAMS

Cincinnati, Ohio
February, 1974

CONTENTS

CONTENTS

LINEAR PROGRAMMING
FOR
DECISION MAKING

An Applications Approach

*

1

The Problem

Linear programming is a mathematical technique that has been developed to help managers make decisions. Instead of presenting a formal definition of a linear program at this time, let us begin our discussion by presenting some typical problems where linear programming can be used.

(1) A manufacturer wants to develop a production schedule and an inventory policy that will satisfy sales demands in future periods. Ideally the schedule and policy will enable the company to satisfy demand and at the same time *minimize* the total production and inventory costs.

(2) A financial analyst must select an investment portfolio from a variety of stock and bond investment alternatives. The analyst would like to establish the portfolio that *maximizes* the return on investment.

(3) A marketing manager wants to determine how he should allocate a fixed advertising budget among alternative advertising media such as radio, television, newspaper, magazine, etc. The manager would like to determine the media schedule that *maximizes* the advertising effectiveness.

(4) A company has warehouses in a number of locations throughout the United States that are intended to serve its

many markets. Given a set of customer demands for its products, the company would like to determine what warehouse should ship how much product to what customers such that the total transportation costs are *minimized*.

Although these are but a few of the possible applications where linear programming has been used successfully, the examples do point out the broad nature of the types of problems that can be tackled using linear programming. Even though the applications are diverse, a close scrutiny of the examples points out one basic property that all of these problems have in common. That is, in each example problem we were concerned with *maximizing* or *minimizing* some quantity. In example 1 we wanted to minimize costs, in example 2 we wanted to maximize return on investment, in example 3 we wanted to maximize total advertising effectiveness, and in example 4 we wanted to minimize total transportation costs. In linear programming terminology the maximization of a quantity or the minimization of a quantity is referred to as the *objective* of the problem. Thus, the objective of all linear programs is to maximize or minimize some quantity.

A second property that is common to all linear programming problems is that there are restrictions or *constraints* that limit the degree to which we can pursue our objective. In example 1 the manufacturer is restricted in terms of how far he can reduce costs by the constraints that guarantee that product demand be satisfied and by the constraints that indicate that production capacities are limited. The financial analyst's portfolio problem is constrained by the total amount of investment funds available and the maximum amounts that can be invested in each stock or bond. The marketing manager's media selection decision is constrained by a fixed advertising budget and the availability of the various media. In the transportation problem the minimum cost shipping schedule is constrained by the supply of product available at each warehouse. Thus, constraints are another general feature of every linear programming problem.

Although we have yet to define linear programming in a formal manner, we have nonetheless been able to talk about some typical problems where linear programming has been applied, and as a result we have been able to recognize two properties that are common to all linear programs: the objective and the constraints. In the following discussion, we will illustrate how linear programming can be used to solve problems of the above nature. In the process, we will also arrive at a formal definition of linear programming.

1.1 A Simple Problem

Let us consider the problem currently being analyzed by the management of Par, Inc., a small manufacturer of golf equipment and supplies. Par has been convinced by its distributor that there is an existing market for both a medium and a high priced golf bag. In fact, the distributor is so confident of the market that if Par can make the bags at a competitive price, the distributor has agreed to purchase everything Par can manufacture over the next three months.

After a thorough investigation of the steps involved to manufacture a golf bag, Par has determined that each golf bag produced will require the following operations:

1. cutting and dyeing of material,
2. sewing,
3. finishing (e.g., inserting umbrella holder, club separators, etc.),
4. inspection and packaging.

The head of manufacturing has analyzed each of the operations and has concluded that if the company produces a medium priced Standard model, each bag produced will require $7/10$ of an hour in the cutting and dyeing department, $1/2$ of an hour in the sewing department, 1 hour in the finishing department, and $1/10$ of an hour in the inspection and packaging department. Similarly, the more expensive Deluxe model will require 1 hour of cutting and dyeing time, $5/6$ of an hour of sewing time, $2/3$ of an hour of finishing time, and $1/4$ of an hour of inspection and packaging time. This production information is summarized in Table 1–A.

The accounting department has analyzed these production figures, assigned all relevant variable costs, and has arrived at prices for both bags that will result in a profit of $10 for every Standard bag produced and $9 for every Deluxe bag produced.

Departments

Product	Cutting & Dyeing	Sewing	Finish- ing	Inspection & Packaging
Standard Bag	7/10 hr.	1/2 hr.	1 hr.	1/10 hr.
Deluxe Bag	1 hr.	5/6 hr.	2/3 hr.	1/4 hr.

Table 1–A. Production Operations and Production Requirements Per Bag

In addition, the head of manufacturing has studied his work load for the next three months and estimates that he should have available a maximum of 630 hours of cutting and dyeing time, 600 hours of sewing time, 708 hours of finishing time, and 135 hours of inspection and packaging time. Par's problem then is to determine how many Standard and how many Deluxe bags should be produced in order to *maximize profit*. If you were in charge of production scheduling for Par, Inc., what decision would you make given the above information? That is, how many Standard and how many Deluxe bags would you produce in the next three months? Write your decision in the spaces below. Later you can check and see how good your decision was.

Number of Standard Bags	Number of Deluxe Bags	Total Profit

1.2   The Objective Function

As we pointed out earlier, every linear programming problem has a specific objective. For the Par, Inc. problem, the objective is to maximize profit. We can write this objective in a more specific form with the introduction of some simple notation. Let

x_1 = the number of Standard bags Par, Inc. produces,

and

x_2 = the number of Deluxe bags Par, Inc. produces.

Then Par's profit will be made up of two parts: (1) the profit made by producing x_1 Standard bags and (2) the profit made by producing x_2 Deluxe bags. Since Par makes \$10 for every Standard bag produced, the company will make \$$10x_1$ if x_1 Standard bags are produced. Also, since Par makes \$9 for every Deluxe bag produced, the company will make \$$9x_2$ if x_2 Deluxe bags are produced. If we denote the total profit with the symbol z, then in terms of our notation

$$\text{total profit} = z = \$10x_1 + \$9x_2.$$

From now on we will just assume that the profit is measured in dollars, and we will write our expression for total profit without the dollar signs. That is,

$$\text{total profit} = z = 10x_1 + 9x_2. \tag{1.1}$$

The solution to Par, Inc.'s problem, then, is to make the *decision* that will maximize total profit. That is, Par, Inc. must determine the values of the *variables* x_1 and x_2 that will yield the highest possible value of z. In linear programming terminology we refer to x_1 and x_2 as the *decision variables*. Since the objective —maximize total profit—is a *function* of these decision variables, we refer to $10x_1 + 9x_2$ as the *objective function*. Thus, in linear programming terminology, we say that Par's goal or objective is to maximize the value of its objective function. Using max as an abbreviation for maximize, we can now write our objective as follows:

$$\max z = \max 10x_1 + 9x_2. \tag{1.2}$$

Suppose Par decided to make 400 Standard bags and 200 Deluxe bags. What would the profit be for this particular production combination? In terms of our decision variables x_1 and x_2, this production combination would mean that

$$x_1 = 400, \text{ and } x_2 = 200.$$

The corresponding profit would be

$$z = 10(400) + 9(200)$$
$$= 4000 + 1800$$
$$= 5800.$$

What if Par decided upon a different production combination such as producing 800 Standard bags and no Deluxe bags. In this case, Par's profit would be

$$z = 10(800) + 9(0)$$
$$= 8000.$$

Certainly the latter production combination is a better production combination for Par, Inc. in terms of the stated objective of maximizing profit. However, it may not be possible for Par, Inc. to manufacture 800 Standard bags and no Deluxe bags. Let us look at the number of hours that will be required for each of the four operations if we consider this particular production combination. Using the data in Table 1–A you can see that this particular product combination would require 560 hours of cutting and dyeing time, 400 hours of sewing time, 800 hours of finishing time, and 80 hours of inspection and packaging time. Can Par, Inc. produce 800 Standard bags? The answer is no because one department—the finishing department—does not have a sufficient number of hours available. Thus because of the constraints on the number of hours available, Par, Inc. is not able to consider 800 Standard bags and no Deluxe bags as an acceptable production alternative. In fact Par, Inc. can only consider the production alternatives that have total hour requirements less than or equal to the maximum hours available for each of the four operations.

In the Par, Inc. problem any particular production combination of Standard and Deluxe bags is referred to as a *solution* to the problem; however, only those solutions which satisfy *all* of the constraints of the problem are referred to as *feasible solutions*. The particular feasible production combination or feasible solution that results in the largest profit will be referred to as the *optimal* production combination or equivalently the *optimal solution* to the problem. At this point, however, we have no idea what the optimal solution will be because we have not developed a procedure for identifying feasible solutions. The procedure for determining the feasible solutions requires us to first identify all the constraints of the problem.

1.3   The Constraints

We know that every bag has to go through four manufacturing operations. Since there is a limited amount of production time available for each of these operations, we can expect four restrictions or constraints that will limit the total number of golf bags Par can produce. Hence, the next step in our linear programming approach to this problem will be to clearly specify all of the constraints associated with the problem.

Cutting and Dyeing Capacity Constraint

From our production information (see Table 1–A) we know that every Standard bag Par manufactures will use $7/10$ of an hour of cutting and dyeing time. Hence, the total number of hours of cutting and dyeing time used in the manufacture of x_1 Standard bags will be $7/10 x_1$. On the other hand, every Deluxe bag Par produces will use 1 hour of cutting and dyeing time, so x_2 Deluxe bags will use up $1x_2$ hours of cutting and dyeing time. The total cutting and dyeing time required for the production of x_1 Standard bags and x_2 Deluxe bags is then given by

$$\text{total cutting and dyeing time required} = 7/10 x_1 + 1x_2.$$

Since the head of manufacturing has stated that Par has at most 630 hours of cutting and dyeing time available, it follows that the product combination we select must satisfy the requirement

$$7/10 x_1 + 1x_2 \leq 630, \tag{1.3}$$

where the symbol $\leq$ means *less-than-or-equal-to*. Relationship (1.3) is referred to as an inequality and denotes the fact that the total number of hours used for the cutting and dyeing operation in the production of x_1 Standard bags and x_2 Deluxe bags must be less than or equal to the maximum amount of cutting and dyeing time Par, Inc. has available. Inequality (1.3), then, represents the cutting and dyeing constraint for Par, Inc.

Sewing Capacity Constraint

We know that every Standard bag manufactured will require $1/2$ of an hour of sewing time, and every Deluxe bag manufactured will require $5/6$ of an hour of sewing time. Since there are 600 hours of sewing time available, it follows that

$$1/2 x_1 + 5/6 x_2 \leq 600. \tag{1.4}$$

Inequality (1.4) is the mathematical representation of the sewing constraint.

Finishing Capacity Constraint

Every Standard bag produced will require 1 hour of finishing time, and every Deluxe bag produced will require ⅔ of an hour of finishing time. Since there are 708 hours of finishing time available, we may represent this constraint mathematically as

$$1x_1 + \tfrac{2}{3}x_2 \leq 708. \tag{1.5}$$

Inspection and Packaging Constraint

For the inspection and packaging operation, we have 135 hours of time available. Since every Standard bag requires ¹⁄₁₀ of an hour and every Deluxe bag requires ¼ of an hour, we must require that

$$\tfrac{1}{10}x_1 + \tfrac{1}{4}x_2 \leq 135.$$

We now have specified the mathematical relationships for the constraints associated with our four production operations. Are there any other constraints we may have forgotten? Can Par, Inc. produce a negative number of Standard or Deluxe bags? Clearly, the answer is no! Thus, in order to prevent our decision variables x_1 and x_2 from having negative values we must add two additional constraints

$$x_1 \geq 0 \text{ and } x_2 \geq 0, \tag{1.7}$$

where the symbol $\geq$ means *greater-than-or-equal-to*. These constraints ensure that the solution to our problem will contain nonnegative values and are thus referred to as the *nonnegativity constraints*. These nonnegativity constraints, which require *all* decision variables to be greater than or equal to zero, are a general feature of all linear programming problems and will be written in the following abbreviated form:

$$x_1,\ x_2 \geq 0. \tag{1.7}$$

1.4   The Mathematical Statement of the Par, Inc. Problem

The mathematical statement or mathematical formulation of our Par, Inc. problem is now complete. That is, we have succeeded in translating the objective and constraints of the "real-world" problem into a set of mathematical relationships which we refer to as a *mathematical model*. We can now write the complete mathematical model for the Par, Inc. problem as follows:

max $10x_1 + 9x_2$

s.t. (an abbreviation for subject to)

$\frac{7}{10}x_1 + 1x_2 \leq 630$ Cutting and Dyeing

$\frac{1}{2}x_1 + \frac{5}{6}x_2 \leq 600$ Sewing

$1x_1 + \frac{2}{3}x_2 \leq 708$ Finishing

$\frac{1}{10}x_1 + \frac{1}{4}x_2 \leq 135$ Inspection and Packaging

$x_1, x_2 \geq 0$

Our job now is to find that product mix (i.e., the combination of x_1 and x_2) which satisfies all of the constraints and, at the same time, yields a value for the objective function that is greater than or equal to the value given by any other feasible solution. Once we have done this, we will have found the optimal solution to our problem.

The above mathematical model of the Par, Inc. problem is what is known as a *linear program*. You can see that the problem has the objective and constraints that we said earlier were common properties of all linear programs. But what is the special feature of this mathematical model that makes us want to call it a *linear* program? The special feature of this model that makes it a linear program is that the objective function is a linear function of the decision variables and the constraint functions (the left-hand sides of the constraint inequalities) are also linear functions of the decision variables.

Mathematically speaking we refer to functions where each of the variables appears in a separate term and is raised to the first power as linear functions. Our objective function $(10x_1 + 9x_2)$ is linear since each decision variable appears in a separate term and has an exponent of one. If the objective function had appeared as $(10x_1^2 + 9\sqrt{x_2})$ it would not have been a linear function and we would not have had a linear program. The amount of production time required in the cutting and dyeing department $(\frac{7}{10}x_1 + 1x_2)$ is also a linear function of the decision variables for the same reasons. Similarly, the functions on the left-hand side of all the constraining inequalities (the constraint functions) are linear functions. Thus, we see that our mathematical formulation of the Par, Inc. problem is indeed a linear program.

An important practical question now comes to mind. What are the properties of the real-world problem faced by Par, Inc., or of real-world problems in general, that make it possible for us to formulate a linear programming model of the problem? There are two conditions or assumptions that must be satisfied before a real-world problem can be formulated and solved as a linear program. These are *proportionality* and *additivity*.

The proportionality assumption means that the value of each term in the objective function and in the constraint functions is proportional to the value of the decision variable appearing in that term. This assumption is satisfied for the Par, Inc. problem. For example, if we doubled the production of Standard bags from 100 to 200 the profit resulting from the sale of Standard bags would increase from $1000 to $2000. The amount of production time devoted to Standard bags would also double in each department. The cutting and dyeing time required would increase from 70 to 140 hours; the sewing time from 50 to 100 hours; the finishing time from 100 to 200 hours; and the inspection and packaging time from 10 to 20 hours. Similarly, any change in the production level for the Deluxe bag would lead to proportionate changes in the objective function and each of the constraint functions.

The additivity assumption means that the total value of the objective function must be equal to the sum of the terms resulting from each decision variable. Also, the total value of the constraint functions must be equal to the sum of the terms resulting from each of the decision variables. This additivity assumption is satisfied for our Par, Inc. problem since the objective function $(10x_1 + 9x_2)$ is the sum of the profits resulting from Standard and Deluxe bags respectively, and the total time used in each department is the sum of the times required for Standard and Deluxe bags respectively.

The reason these assumptions of proportionality and additivity are so important is that if these assumptions are satisfied by a particular real-world problem then a linear objective function and linear constraint functions do provide an accurate mathematical representation of the corresponding real-world relationships. As we saw earlier, a linear program can be characterized as a problem involving a linear objective function and linear constraint functions. Thus if these assumptions of proportionality and additivity are satisfied, a linear programming model provides an accurate mathematical representation of the real-world problem.

We must always remember that a linear programming model is only a mathematical representation of the real-world problem. How good the representation is will depend to a large extent upon how closely the proportionality and additivity assumptions represent the actual conditions in the real-world situation. The closer these assumptions come to being satisfied, the closer the linear programming model will come to being an accurate mathematical representation of the real-world problem. If these assumptions are satisfied (or nearly satisfied), then we can be reasonably sure that the optimal solution to our linear program will indicate the optimal decision for our real-world problem.

In the remainder of this text we will be assuming that the proportionality and additivity assumptions are acceptable and hence that a linear programming model is appropriate for finding optimal solutions.

1.5   The Graphical Solution Approach

An easy way to solve a simple linear programming problem that contains only two decision variables is the graphical solution procedure. Although the graphical method is quite awkward for solving three-variable problems, and cannot be used for larger problems, the insight gained from studying this method will be invaluable as an aid to understanding some of the more difficult concepts discussed later in the book. In addition, the graphical method provides an intuitive basis for more practical solution methods such as the Simplex method which we will discuss in Chapter 2.

Let us begin our graphical solution procedure by developing a graph that can be used to display the possible solutions (x_1 and x_2 values) for the Par, Inc. problem. On our graph (Figure 1–A), the values of x_1 will be shown on the horizontal axis, and the values of x_2 will be shown on the vertical axis. Any point on the graph can be identified by the x_1 and x_2 values which indicate the position of the point along the x_1 and x_2 axes respectively. Since every point (x_1, x_2) corresponds to a possible solution, every point on our graph is called a *solution point*. The solution point where $x_1 = 0$ and $x_2 = 0$ will be referred to as the origin of the graph.

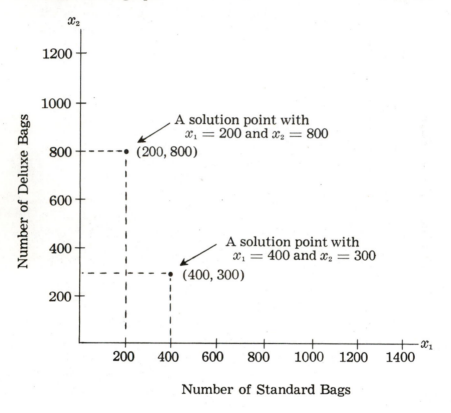

Figure 1–A. Graph of Solution Points for the Two-Variable Par, Inc. Problem

Our next step will be to show graphically which of the possible combinations of x_1 and x_2—solution points—correspond to feasible solutions to our linear program. Since in our linear programming problem both x_1 and x_2 must be nonnegative, we need only consider points where $x_1 \geq 0$ and $x_2 \geq 0$. This is indicated in Figure 1–B by arrows pointing in the direction of production combinations that will satisfy the nonnegativity relationships in the problem. In all future graphs, we will assume that the nonnegativity relationships hold, and hence we will only draw the portion of the graph corresponding to nonnegative x_1 and x_2 values.

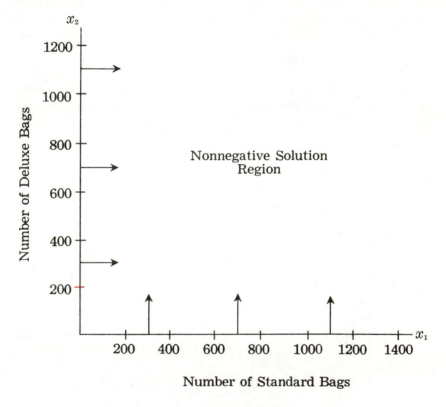

Figure 1–B.   The Nonnegativity Constraints

Previously, we saw that the inequality representing the cutting and dyeing constraint was of the form

$$\tfrac{7}{10}x_1 + 1x_2 \le 630.$$

We can show all solution points that satisfy this relationship by first graphing the line corresponding to the equation

$$\tfrac{7}{10}x_1 + 1x_2 = 630.$$

The graph of this equation is found by identifying two points that lie on the line and then drawing a line through the points. Setting $x_1 = 0$ and solving for x_2 we see that the point ($x_1 = 0$, $x_2 = 630$) satisfies the above equation. To find a second point satisfying this equation we set $x_2 = 0$ and solve for x_1. Doing this we get $\frac{7}{10}x_1 + 1(0) = 630$, or $x_1 = 900$. Thus, a second point satisfying the equation is ($x_1 = 900$, $x_2 = 0$). Given these two points we may now graph the line corresponding to the equation

$$\tfrac{7}{10}x_1 + 1x_2 = 630.$$

This line, which will be called the cutting and dyeing *constraint line*, is shown in Figure 1–C. For purposes of identification, we

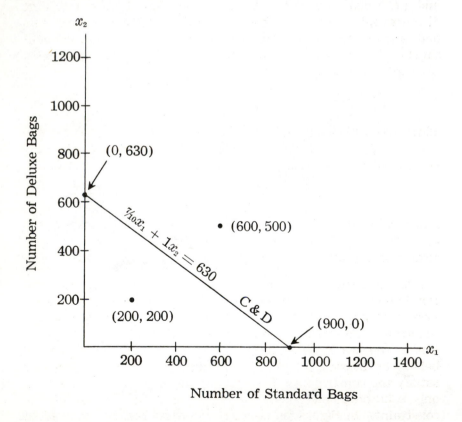

Figure 1–C.   The Cutting and Dyeing Constraint Line

shall label this line C & D to indicate that it represents the cutting and dyeing constraint.

Recall that the inequality representing the cutting and dyeing constraint is

$$\tfrac{7}{10}x_1 + 1x_2 \leq 630.$$

Can you identify all of the solution points that satisfy this constraint? Well, since we have the line where $\tfrac{7}{10}x_1 + 1x_2 = 630$, we know any point on this line must satisfy the constraint. But where are the solution points satisfying $\tfrac{7}{10}x_1 + 1x_2 < 630$? Let us consider two solution points: $(x_1 = 200, x_2 = 200)$ and $(x_1 = 600, x_2 = 500)$. You can see from Figure 1–C that the first solution point is below the constraint line and the second is above the constraint line. Which of these solutions will satisfy the cutting and dyeing constraint? For the point $(x_1 = 200, x_2 = 200)$ we see that

$$\tfrac{7}{10}x_1 + 1x_2 = \tfrac{7}{10}(200) + 1(200) = 340.$$

Since the 340 hours is less than the 630 hours available, the $x_1 = 200$, $x_2 = 200$ production combination or solution point satisfies the constraint. For $x_1 = 600$, $x_2 = 500$ we have

$$\tfrac{7}{10}x_1 + 1x_2 = \tfrac{7}{10}(600) + 1(500) = 920.$$

Since the 920 hours is greater than the 630 hours available, the $x_1 = 600$, $x_2 = 500$ solution point does not satisfy the constraint and is thus an unacceptable production alternative.

Are you ready to answer the question of where are the solution points that satisfy the cutting and dyeing constraint? Your answer should be that any point *below* the cutting and dyeing constraint line satisfies the constraint. You may want to prove this to yourself by selecting additional solution points above and below the constraint line and checking to see if the solutions satisfy the constraints. You will see that for the $\leq$ constraints only solution points on or below the constraint line satisfy the constraint. In Figure 1–D we indicate all such points by shading in the region of the graph corresponding to the solution points that satisfy the cutting and dyeing constraint.

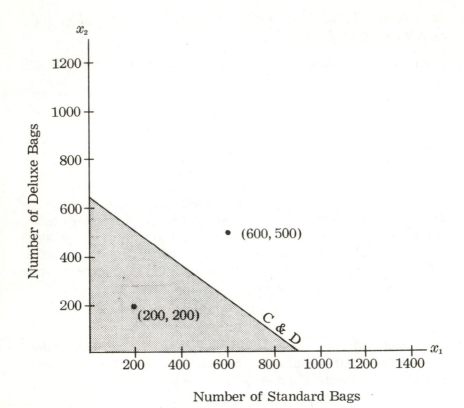

Figure 1–D.   Feasible Region for the Cutting and
             Dyeing Constraint

Next let us identify all solution points that satisfy the sewing
constraint

$\frac{1}{2}x_1 + \frac{5}{6}x_2 \leq 600.$

We start by drawing the constraint line corresponding to the
equation

$\frac{1}{2}x_1 + \frac{5}{6}x_2 = 600.$

As before, the graphing of a line is most easily done by finding
two points on the line and then connecting them. Thus, we first
set x_1 equal to zero and solve for x_2, which yields the point

$(x_1 = 0, x_2 = 720)$. Next, we set x_2 equal to zero and solve for x_1, which gives the second point $(x_1 = 1200, x_2 = 0)$. In Figure 1–E we have drawn the line corresponding to the sewing constraint. For identification purposes, we will label this line S.

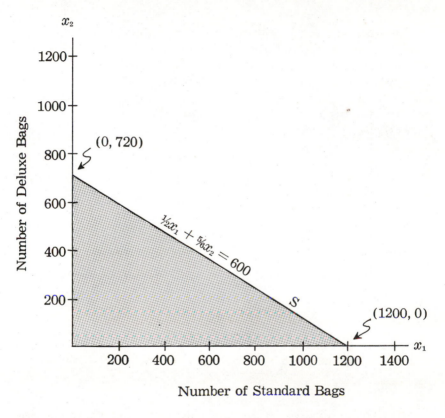

Figure 1–E.   Feasible Region for the Sewing Constraint

Using the same approach we used for the cutting and dyeing constraint, we realize that only points on or below the line will satisfy the sewing time constraint. Thus, in Figure 1–E we have shaded the region corresponding to all feasible production combinations or feasible solution points for the sewing operation.

In a similar manner, we can determine the set of all feasible production combinations for each of the remaining constraints. This has been done, and the results are shown in Figures 1–F and 1–G.   For practice, try to graph the feasible solution region for the finishing (F) constraint and the inspection and packaging (I & P) constraint and see if your results agree with those shown in Figures 1–F and 1–G.   If they don't agree, reread section 1.5 *carefully!*

We now have four separate graphs showing the feasible solution points for each of the four constraints.   In a linear programming problem, we need to identify the solution points that satisfy *all* of the constraints *simultaneously*.   To find these solution points, we can draw our four constraints on one graph and observe the region containing the points that do in fact satisfy all of the constraints.

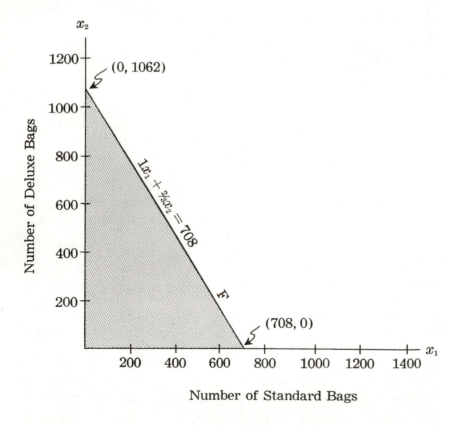

Number of Standard Bags

Figure 1–F.   Feasible Region for the Finishing Constraint

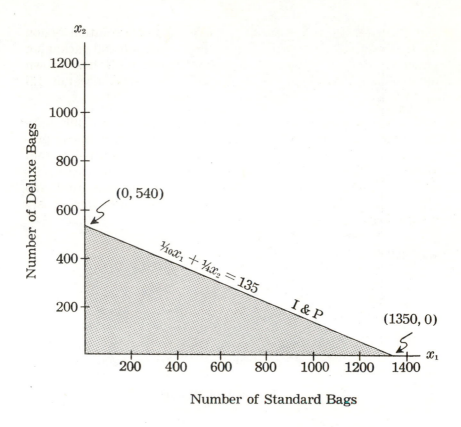

Figure 1–G. Feasible Region for the Inspection and Packaging Constraint

The graphs in Figures 1–D through 1–G can be superimposed to obtain one graph with all four constraints. This combined-constraint graph is shown in Figure 1–H. The shaded region in this figure includes every solution point that satisfies all of the constraints. Since solutions that satisfy all of the constraints are termed *feasible solutions*, the shaded region is called the feasible solution region or simply the *feasible region*. Any point on the boundary of the feasible region or within the feasible region is a *feasible solution point*. You may want to check points outside the feasible region to prove to yourself that these solution points violate one or more of the constraints and are thus infeasible or unacceptable.

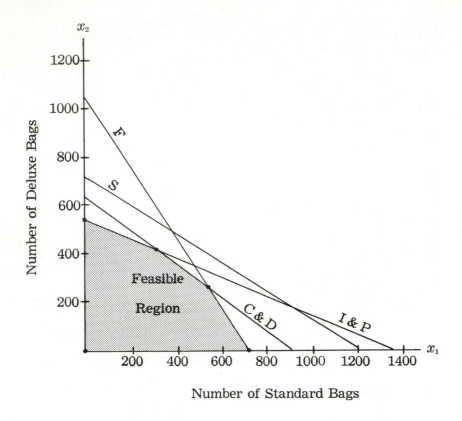

Figure 1–H.   Feasible Solution Region for the Par, Inc. Problem

Now that we have identified the feasible region we are ready to proceed with the graphical solution method and find the optimal solution to the Par, Inc. problem. Recall that the optimal solution for any linear programming problem is the feasible solution that provides the best possible value of the objective function. We could arbitrarily select feasible solution points (x_1, x_2) and compute the associated profit $10x_1 + 9x_2$; however, the difficulty with this approach is that there are too many feasible solutions (actually an infinite number), and thus it would not be possible to evaluate all feasible solutions. Hence this trial-and-error procedure would not guarantee that the optimal solution would be obtained. Thus we would like a systematic way of identifying the feasible solution that does in fact maximize the profit for Par, Inc.

Let us start this final step of our graphical solution procedure by drawing the feasible region on a separate graph. This is shown in Figure 1–I.

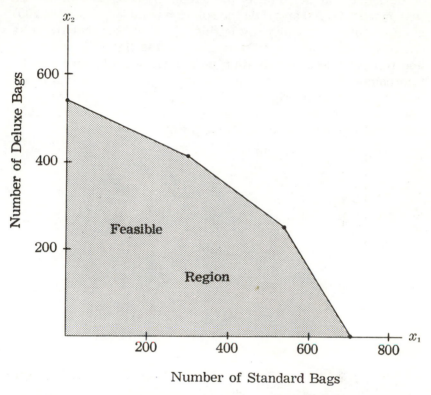

Figure 1–I.   Feasible Solution Region for the Par, Inc. Problem

Rather than selecting an arbitrary feasible solution and computing the associated profit, let us select an arbitrary profit and identify all of the feasible solution points (x_1, x_2) that yield the selected profit. For example, what feasible solution points provide a profit of \$1,800? That is, we are asking what values of x_1 and x_2 in the feasible region will make the objective function

$$10x_1 + 9x_2 = 1800.$$

The above expression is simply the equation of a line; thus all feasible solution points (x_1, x_2) yielding a profit of \$1800 must be on the line. We learned earlier in this section how to graph a constraint line. The procedure for graphing our profit or objective function line is the same. Letting $x_1 = 0$, we see that x_2 must be 200 and thus the solution point ($x_1 = 0$, $x_2 = 200$) is on the line. Similarly, by letting $x_2 = 0$ we see that the solution point ($x_1 = 180$, $x_2 = 0$) is also on the line. Drawing the line through these two solution points will identify all the solution points that have a profit of $1800. A graph of this profit line is presented in Figure 1–J. From this graph you can see that there are an infinite number of feasible production combinations that will provide an $1800 profit.

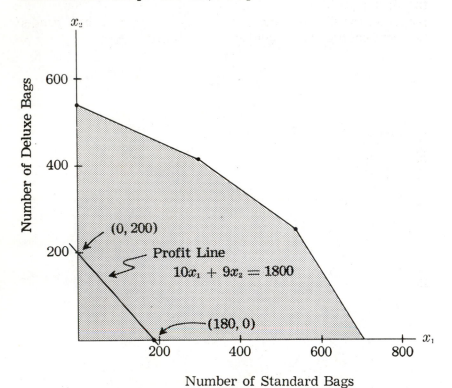

Figure 1–J. $1800 Profit Line for the Par, Inc. Problem

Since our objective is one of finding the feasible solution point that has the highest profit, let us proceed by selecting higher profit values and finding the feasible solution points that yield the stated profits. For example, what solution points provide a

$3600 profit?   What solution points provide a $5400 profit? To answer these questions we must find the x_1 and x_2 values that are on the following lines:

$$10x_1 + 9x_2 = 3600$$

and

$$10x_1 + 9x_2 = 5400.$$

Using our previous procedure for graphing profit and constraint lines, we have drawn the $3600 and $5400 profit lines on the graph in Figure 1–K. While not all solution points on the $5400 profit line are in the feasible region, at least some points on the line are, and thus we can obtain a feasible production combination that provides a $5400 profit.

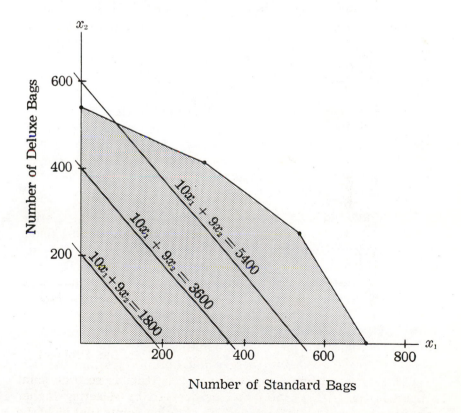

Figure 1–K.   Selected Profit Lines for the Par, Inc. Problem

Can we find a solution yielding even higher profits? Look at Figure 1–K and see what general observations you can make about the profit lines. You should be able to identify the following properties:

1.   the profit lines are *parallel* to each other,
and

2.   higher profit lines occur as we move further from the origin.

These properties, which hold for all two decision variable linear programs, can now be used to determine the optimal solution to our problem. We know that if we continue to move our profit line further from the origin such that it remains parallel to the other profit lines, we can obtain solution points that yield higher and higher values for the objective function. However, at some point we will find that any further outward movement will place the profit line outside the feasible region. Since points outside the feasible region are unacceptable, the point in the feasible region which lies on the highest profit line is the optimal solution to our linear program.

You should now be able to identify the optimal solution point for the Par, Inc. problem. Use a ruler or a straight piece of paper and move the profit line as far from the origin as you can. What is the last point in the feasible region that you reach? This point, which is the optimal solution is shown graphically in Figure 1–L.

The optimal values of the decision variables x_1 and x_2 are the x_1 and x_2 values at the optimal solution point. Depending upon the accuracy of your graph, you may or may not be able to read the *exact* x_1 and x_2 values from the graph. The best we can do with respect to the optimal solution point in Figure 1–L is to conclude that the optimal production combination consists of approximately 550 Standard bags (x_1) and 250 Deluxe bags (x_2). As you will see later, the actual x_1 and x_2 values at the optimal solution point are $x_1 = 540$ and $x_2 = 252$. In section 1.7 we show how to compute these exact optimal solution values.

The optimal solution of 540 Standard bags and 252 Deluxe bags yields a profit of $10(540) + 9(252) = \$7668$. How close did you come to the actual optimal solution with your decision from section 1.1?

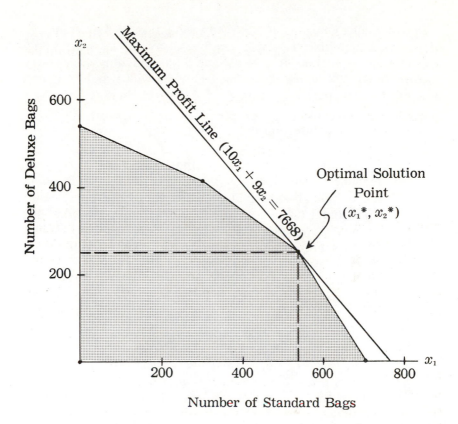

Figure 1–L. Optimal Solution Point for the Par, Inc. Problem

In addition to the optimal solution and the expected profit, the management of Par, Inc. will probably want information about the production time requirements for each production operation. We can determine this information by substituting the optimal x_1 and x_2 values into the constraints of our linear program. For the Par, Inc. problem the production time requirements using equations (1.3) through (1.6) are as follows:

$$\tfrac{7}{10}(540) + 1(252) = 630 \text{ hours of cutting and dyeing time,}$$

$$\tfrac{1}{2}(540) + \tfrac{5}{6}(252) = 480 \text{ hours of sewing time,}$$

$$1(540) + \tfrac{2}{3}(252) = 708 \text{ hours of finishing time,}$$

$$\tfrac{1}{10}(540) + \tfrac{1}{4}(252) = 117 \text{ hours of inspection and packaging time.}$$

We can tell management that the production of 540 Standard bags and 252 Deluxe bags will require all available cutting and dyeing time (630 hours) and all available finishing time (708 hours), while 120 hours of sewing time (600–480) and 18 hours of inspection and packaging time (135–117) remain idle.

Could we have used our graphical analysis to provide some of this production information? The answer is, yes. By finding our optimal solution point on Figure 1–H, we can see that the cutting and dyeing and the finishing constraints restrict or bind our feasible region at this point. Thus, this solution point requires the use of all available time for these two operations. On the other hand, since the sewing and the inspection and packaging constraints are not restricting the feasible region at the optimal solution point, we can expect some unused time for these two operations.

As a final comment on the graphical analysis of the Par, Inc. problem, we call your attention to the sewing capacity constraint as shown in Figure 1–H. Note in particular that this constraint did not affect the feasible region. That is, the feasible region would be the same whether the sewing capacity constraint was included or not. This tells us that there is enough sewing time available to accommodate any production level that can be achieved by the other three departments. Since the sewing constraint does not affect the feasible region and thus cannot affect the optimal solution, it is called a *redundant* constraint. Redundant constraints can be dropped from the problem without having any effect upon the optimal solution.

A Summary of the Graphical Solution Procedure

As you have seen, the graphical solution procedure is one method of solving two-variable linear programming problems such as the Par, Inc. problem. The steps of the graphical solution procedure are outlined below:

1. prepare a graph of the feasible solution points for each of the constraints;

2. determine the feasible solution region by identifying the solution points that satisfy all of the constraints simultaneously;

3. draw a profit line showing all values of the x_1 and x_2 variables that yield a specified value of the objective function;

4. move parallel profit lines away from the origin [1] until further movement would take the profit line completely outside the feasible region;

5. the feasible solution point that is touched by the highest possible profit line is the optimal solution;

6. determine, at least approximately, the optimal values of the decision variables by reading the x_1 and x_2 values at the optimal solution point directly from the graph.

1.6   Extreme Points and the Optimal Solution

Let us suppose that the profit for the Par, Inc. Standard bag is reduced from $10 to $5 per bag, while the profit for the Deluxe bag and all the constraint conditions remain unchanged. Our complete linear programming model of this new problem is identical to the mathematical model in section 1.4 except for the following revised objective function:

$$\max z = \max 5x_1 + 9x_2.$$

How does this change in the objective function affect the optimal solution to our Par, Inc. problem?   In Figure 1–M we show the graphical solution to the Par, Inc. problem with the revised objective function.   Note that since our constraints have not changed, the feasible solution region has not changed. However, our profit lines have been altered to reflect the new objective function.

By moving the profit line in a parallel manner away from the origin, we found the optimal solution point shown in Figure 1–M.   The exact values of the decision variables at this point are $x_1 = 300$ and $x_2 = 420$.   Thus, the reduced profit for the Standard bags has caused us to change our optimal solution.   In fact, as you might have suspected, we are cutting back the production of the lower-profit Standard bags and increasing the production of the higher-profit Deluxe bags.

---

[1] This rule applies specifically to a maximization objective such as found in the Par, Inc. problem.   A modification of this step for minimization problems will be presented in Chapter 3.

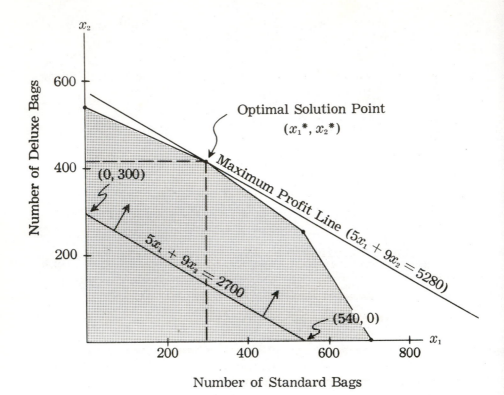

Figure 1–M. Optimal Solution Point for the Par, Inc. Problem
with an Objective Function of $5x_1 + 9x_2$

What have you noticed about the location of the optimal solutions in the two linear programming problems we have solved thus far? Look closely at the graphical solutions in Figures 1–L and 1–M. An important observation that you should be able to make is that the optimal solutions occur at one of the vertices or "corners" of the feasible region. In linear programming terminology these vertices are referred to as the *extreme points* of the feasible region. Thus, the Par, Inc. problem has five vertices or five extreme points for its feasible region (see Figure 1–N). We can now state our observation about the location of optimal solutions as follows:

The optimal solution to every linear programming problem occurs at an extreme point of the feasible solution region for the problem.

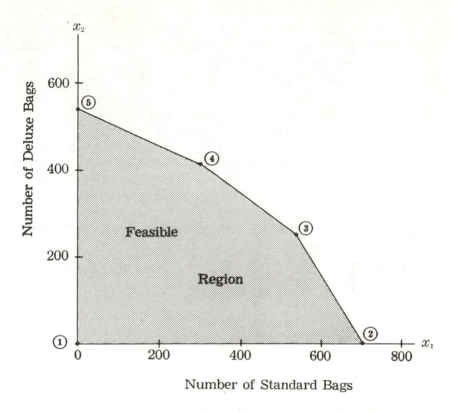

Figure 1–N. The Five Extreme Points of the Feasible Region
for the Par, Inc. Problem

This is a very important property of all linear programming problems because it says that if you are looking for the optimal solution to a linear programming problem, you do not have to evaluate all feasible solution points. In fact, you *only* have to consider the feasible solutions that occur at the extreme points of the feasible region. Thus, for the Par, Inc. problem, instead of computing and comparing the profit for all feasible solutions, we know we can find the optimal solution for the problem by evaluating the five extreme point solutions and selecting the one that provides the highest profit. Actually, the graphical solution procedure is nothing more than a convenient way of identifying the optimal extreme point for two-variable problems.

To help convince yourself that the optimal solution to a linear program always occurs at an exteme point, select sev-

eral different objective functions for the Par, Inc. problem and
graphically find the optimal solution for each case. You will
see that as you move the profit lines away from the origin, the
last feasible solution point—the optimal solution point—is always one of the extreme points.

What happens if the highest profit line coincides with one
of the constraint lines on the boundary of the feasible region?
This case is shown for a $4x_1 + 10x_2$ objective function in Figure 1–O. Does the optimal solution still occur at an extreme
point? The answer is yes. In fact, for this case the optimal
solution occurs at extreme point ④, extreme point ⑤, and any
solution point on the line joining these two extreme points.
This is the special case of alternate optimal solutions which

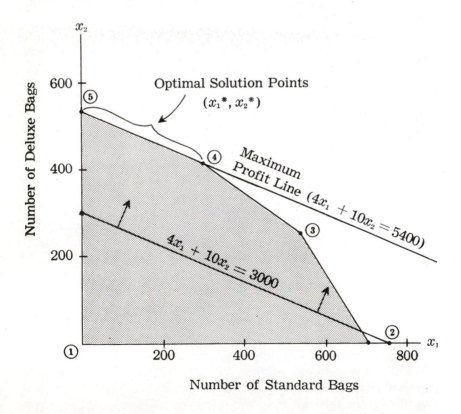

Figure 1–O.   Optimal Solution Points for the Par, Inc. Problem
with an Objective Function of $4x_1 + 10x_2$

will be discussed in detail in Chapter 3. Here we are merely concerned with noting that if alternate optimal solutions exist, one optimal solution to the linear programming problem can still always be found by evaluating the extreme point solutions.

1.7 Simultaneous Linear Equations

The graphical solution method is a convenient way to find optimal extreme point solutions for two-variable, linear programming problems; however, for problems having more than two decision variables, this method is unacceptable. In Chapter 2 we will introduce a general algebraic solution procedure, the Simplex method, which can be used to find optimal extreme point solutions for linear programming problems having anywhere from two to several thousand decision variables. The mathematical steps of this Simplex method are based in part upon concepts from the area of simultaneous linear equations. Thus in order to prepare you for using the Simplex method and understanding how the method identifies optimal extreme point solutions, we must briefly study simultaneous linear equations. We shall find that knowledge about simultaneous linear equations is also helpful in identifying the exact x_1 and x_2 values at the optimal extreme point in a graphical solution of a two-variable problem.

If we have a system of two or more linear equations involving the variables x_1, x_2, x_3, etc., then we say we have a system of simultaneous linear equations. For example, the following is a system of two simultaneous linear equations involving two variables.

$$2x_1 + 1x_2 = 16 \tag{1.8}$$
$$-1x_1 + \tfrac{1}{2}x_2 = 2 \tag{1.9}$$

A solution to a system of simultaneous linear equations is any combination of values for the variables x_1, x_2, x_3, etc., that satisfies all the equations at the same time. The solution to the above system of simultaneous equations is $x_1 = 3$, $x_2 = 10$, as you can easily verify by substituting these values into (1.8) and (1.9). The way in which these solution values were found was by the use of elementary row operations which we will now discuss.

Elementary Row Operations

A solution to a set of simultaneous linear equations can be found by applying a sequence of the following two *elementary row operations:*

1. multiplying a row or equation by a non-zero number; and
2. adding a non-zero number times one row or equation to another row or equation.

These elementary row (equation) operations will change the coefficients of the variables in the linear equations; however, it is important to realize that although the coefficients of the variables may change, *the elementary row operations do not change the solution (x_1, x_2, x_3, etc.) to the set of linear equations.* That is, while the linear equations may appear in a different form after applying the elementary row operations, the same values of the variables (x_1, x_2, x_3, etc.) solve them. We will show how these elementary row operations can be used to obtain the solution to a set of linear equations after we have given examples of how the elementary row operations can change the coefficients of the variables in an equation.

Let us return to the two-variable simultaneous linear equations (1.8) and (1.9). Operation 1 states that we can multiply any row or equation by a non-zero value and the solution will not be changed. To illustrate this operation we multiply equation (1.8) by 3 and equation (1.9) by -2. The resulting set of linear equations are:

$$6x_1 + 3x_2 = 48 \tag{1.10}$$
$$2x_1 - 1x_2 = -4 \tag{1.11}$$

While the coefficients of the variables and the right-hand sides of the equations have changed, the solution to the equations has not. That is, $x_1 = 3$ and $x_2 = 10$ satisfy equations (1.10) and (1.11) as well as the original set of simultaneous linear equations (1.8) and (1.9).

To illustrate elementary row operation 2, let us multiply equation (1.9) by -4 and add the resulting equation to (1.8). Doing this we have

$$-4 \ (-1x_1 + \tfrac{1}{2}x_2 = 2) \rightarrow 4x_1 - 2x_2 = -8$$

added to equation (1.8)
$$\frac{2x_1 + 1x_2 = 16}{6x_1 - 1x_2 = \ \ 8.} \tag{1.12}$$
yields

Leaving equation (1.9) unchanged, we have the following set of linear equations

$$6x_1 - 1x_2 = 8 \qquad\qquad (1.12)$$
$$-1x_1 + \tfrac{1}{2}x_2 = 2. \qquad\qquad (1.9)$$

Again, you can see that $x_1 = 3$ and $x_2 = 10$ still solves the system of equations. The important feature of elementary row operations is that even though they change the coefficients of the linear equations, the solution to the resulting system is the same as the solution to the original system.

Solving a Set of Linear Equations by Elementary Row Operations

We will now show you how elementary row operations can be used to put the linear equations in a form that will allow us to easily determine the solution. Consider again the simultaneous linear equations

$$2x_1 + 1x_2 = 16 \qquad\qquad (1.8)$$
$$-1x_1 + \tfrac{1}{2}x_2 = 2. \qquad\qquad (1.9)$$

We know the elementary row operations can be used to change the coefficients of the variables without changing the value of the solution. Therefore, suppose we could get the left-hand side of equation (1.8) into the form

$$1x_1 + 0x_2$$

and the left-hand side of equation (1.9) into the form

$$0x_1 + 1x_2$$

by performing row operations.

The solution to the set of linear equations could be easily determined because the right-hand side value of the first equation would be the value of x_1 and the right-hand side value of the second equation would be the value of x_2. Let us see what elementary row operations must be performed in order to put our linear equations in the above form.

Multiplying equation (1.8) by ½ yields

$$1x_1 + \tfrac{1}{2}x_2 = 8. \tag{1.13}$$

Adding 1 times this equation to equation (1.9) we have

$$0x_1 + 1x_2 = 10. \tag{1.14}$$

This is the exact form we want for the second equation. To get our first equation in the desired form we can multiply equation (1.14) by $-\tfrac{1}{2}$ and add the resulting equation to equation (1.13). Doing this we obtain

$$1x_1 + 0x_2 = 3. \tag{1.15}$$

Equations (1.14) and (1.15) are now in the form that easily provides the solution $x_1 = 3$ and $x_2 = 10$.

Thus, elementary row operations can be used to change the coefficients of the variables so that the solution can be easily obtained. The exact sequence of elementary row operations is up to you. A variety of sequences can be used to obtain the solution for the equations.

Finding the Exact Location of Graphical Solution Extreme Points

Let us consider extreme point ③ of the Par, Inc. problem as shown in Figure 1–P. In our graphical solution procedure of section 1.5 we identified extreme point ③ as the optimal solution to the original Par, Inc. problem. However, we had difficulty reading the exact values of x_1 and x_2 at extreme point ③ directly from the graph. Actually, the best we could do was to arrive at approximate values for the decision variables.

Referring to Figure 1–P, can you see a way we might find the values of x_1 and x_2 at extreme point ③ without having to read the values directly from the graph? What constraint lines determine the exact location of extreme point ③? You should be able to see that the cutting and dyeing constraint line and the finishing constraint line combine to determine this extreme point. That is, extreme point ③ is on both the cutting and dyeing constraint line

$$\tfrac{7}{10}x_1 + 1x_2 = 630 \tag{1.16}$$

and the finishing constraint line

$$1x_1 + \tfrac{2}{3}x_2 = 708. \tag{1.17}$$

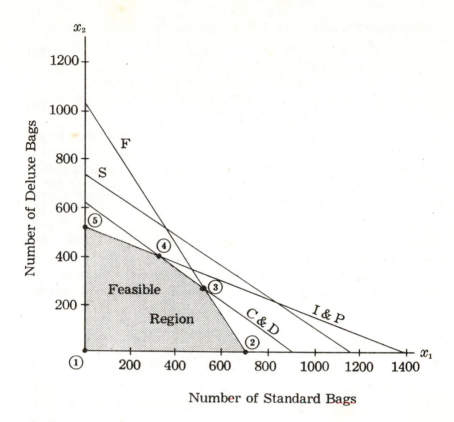

Figure 1–P. Feasible Solution Region for the Par, Inc. Problem

Thus, the values of the decision variables x_1 and x_2 at extreme point ③ must satisfy the simultaneous linear equations given by (1.16) and (1.17).

We can use our elementary row operations on equations (1.16) and (1.17) in order to find the solution. Multiplying equation (1.16) by $10/7$ yields

$$1x_1 + 10/7 x_2 = 900. \tag{1.18}$$

Multiplying equation (1.18) by -1 and adding the result to equation (1.17) gives

$$0x_1 - 16/21 x_2 = -192. \tag{1.19}$$

Multiplying equation (1.19) by $-2\frac{1}{16}$ gives the following desired equation form

$$0x_1 + 1x_2 = 252. \tag{1.20}$$

Multiplying equation (1.20) by $-1\frac{9}{7}$ and adding the result to equation (1.18) provides the other desired equation form

$$1x_1 + 0x_2 = 540. \tag{1.21}$$

You can now see that the x_1 and x_2 values at the extreme point defined by the cutting and dyeing constraint line and the finishing constraint line must be exactly $x_1 = 540$ and $x_2 = 252$.

We have now seen how the solution of simultaneous linear equations can enable us to find exact extreme point solutions when the values cannot be accurately determined from the graph. Problem 12 at the end of this chapter asks you to use the simultaneous linear equation procedure to show that the exact solution at extreme point ④ is $x_1 = 300$ and $x_2 = 420$.

Properties of Solutions to Simultaneous Linear Equations

A solution to a set of simultaneous linear equations is a specific combination of values for the variables that satisfies all of the equations at the same time. Since the algebraic solution procedure for linear programming problems—the Simplex method—involves solving simultaneous linear equations, it will be desirable for us to know something about the number of solutions that exist for a set of simultaneous linear equations.

When using the Simplex method we will need to find the solution, or solutions, to the following two types of systems of linear equations:

1. linear equations having more variables (n) than equations (m) (i. e., n > m) ;
2. linear equations having the same number of variables (n) as equations (m) (i. e., n = m).

When applying the Simplex method to these two types of systems we will always obtain (1) an infinite number of solutions when there are more variables than equations (n > m) or (2) a unique solution when the number of variables is equal to the number of equations.

Consider the following 3-variable, 2-equation set of linear equations:

$$1x_1 + 4x_2 - 1x_3 = 12 \qquad (1.22)$$
$$2x_1 + 1x_2 + \tfrac{3}{2}x_3 = 10. \qquad (1.23)$$

Using our elementary row operations, we can put this system of equations into the following form:

$$1x_1 + 0x_2 + 1x_3 = 4 \qquad (1.24)$$
$$0x_1 + 1x_2 - \tfrac{1}{2}x_3 = 2. \qquad (1.25)$$

If we solve the first equation for x_1 and the second equation for x_2, we can write

$$x_1 = 4 - 1x_3$$

and

$$x_2 = 2 + \tfrac{1}{2}x_3.$$

As you can see, we may select *any* value for x_3 and then use the above equations to determine the values of x_1 and x_2 that will satisfy the equations. For example, one solution with $x_3 = 2$ would have $x_1 = 2$ and $x_2 = 3$. Another solution with $x_3 = 6$ would have $x_1 = -2$ and $x_2 = 5$. Check for yourself to see that both of these solutions satisfy our original linear equations (1.22) and (1.23). Since we can select any value for x_3 and then find appropriate solution values for x_1 and x_2, the n > m linear equations (1.22) and (1.23) have an infinite number of solutions.

Now, suppose we fix the value of x_3 at zero in equations (1.22) and (1.23). In effect we have eliminated x_3 from our system of equations and have the following two-variable, two-equation system.

$$1x_1 + 4x_2 = 12 \qquad (1.26)$$
$$2x_1 + 1x_2 = 10. \qquad (1.27)$$

Using our elementary row operations on the above equations, we obtain the following equivalent system

$$1x_1 + 0x_2 = 4$$
$$0x_1 + 1x_2 = 2.$$

Thus, $x_1 = 4$ and $x_2 = 2$ are the only solution values that satisfy equations (1.26) and (1.27).

We emphasize that the two solution cases discussed above are not the only cases that might occur when solving systems of linear equations; however, they are the only cases which will occur when the Simplex method is used to solve linear programs. Actually, general systems of linear equations with any number of variables (n) and any number of equations (m) may have either no solution, a unique solution, or an infinite number of solutions. The properties of solutions to simultaneous linear equations in general are described in detail by Hadley.[1]

1.8   Standard Form

We have stated that the algebraic solution to linear programming problems will involve solving simultaneous linear equations. In fact, we will have one linear equation for each constraint in the linear programming problem. Since linear programming problems almost always contain some constraints with inequality relationships, such as $\leq$ relationships, we will need a standard procedure for converting each inequality constraint to a linear equation or equality ($=$) form. When all of the constraints of a linear programming problem have been written as equality relationships, we say the problem has been written in its *Standard* form. You will see later that we will convert every linear programming problem to its Standard form before beginning the Simplex solution method.

Let us return to the original Par, Inc. problem and see how we can write this problem in its Standard form. The linear programming model of the Par, Inc. problem is shown below:

$$\max\ 10x_1 + 9x_2$$

s.t.

$$\tfrac{7}{10}x_1 + 1x_2 \leq 630$$
$$\tfrac{1}{2}x_1 + \tfrac{5}{6}x_2 \leq 600$$
$$1x_1 + \tfrac{2}{3}x_2 \leq 708$$
$$\tfrac{1}{10}x_1 + \tfrac{1}{4}x_2 \leq 135$$
$$x_1,\ x_2 \geq 0.$$

In order to transform the above formulation into its Standard form, we must write the constraints corresponding to the four

---

[1] G. Hadley, *Linear Algebra* (Reading, Mass.: Addison-Wesley, 1961).

production operations as equalities. Let us begin with the first constraint

$$\tfrac{7}{10}x_1 + 1x_2 \leq 630.$$

To make this constraint an equality, we must add something to the left-hand side to make up for the difference between 630 and ($\tfrac{7}{10}x_1 + 1x_2$). Since the left-hand side of the constraint varies depending on the values of x_1 and x_2, the amount we add to the left-hand side must be a variable. We refer to a variable that is added to the left-hand side of any less-than-or-equal-to constraint as a *slack variable*. Thus by adding a slack variable (s_1) to the left-hand side of the first equation, the cutting and dyeing constraint can be written as the following equation:

$$\tfrac{7}{10}x_1 + 1x_2 + 1s_1 = 630.$$

If our production combination specifies $x_1 = 100$ and $x_2 = 100$, then

$$\tfrac{7}{10}(100) + 1(100) + 1s_1 = 630$$

or

$$s_1 = 460.$$

If $x_1 = 500$ and $x_2 = 200$, then

$$\tfrac{7}{10}(500) + 1(200) + 1s_1 = 630$$

or

$$s_1 = 80.$$

The physical interpretation of the value of slack variable s_1 is that s_1 indicates the remaining cutting and dyeing production time available after producing x_1 units of the Standard bag and x_2 units of the Deluxe bag.

Using s_2 as the slack variable for the sewing constraint, s_3 as the slack variable for the finishing constraint, and s_4 as the slack variable for the inspection and packaging constraint, the remaining three constraints can be written as the following three linear equations:

$$\tfrac{1}{2}x_1 + \tfrac{5}{6}x_2 + 1s_2 \qquad\qquad = 600$$
$$1x_1 + \tfrac{2}{3}x_2 \qquad + 1s_3 \qquad = 708$$
$$\tfrac{1}{10}x_1 + \tfrac{1}{4}x_2 \qquad\qquad + 1s_4 = 135.$$

In each case the value of the slack variable indicates the excess production time for the corresponding operation after x_1 Standard bags and x_2 Deluxe bags have been produced.

Since we have added four new variables to the problem, we have to be sure to include these variables in the objective function. Since the excess time available for any operation makes no contribution to profit, the profit coefficient associated with each slack variable will be zero. Thus, the Par, Inc. linear programming problem can now be written in Standard form.

$$\max \ 10x_1 + 9x_2 + 0s_1 + 0s_2 + 0s_3 + 0s_4$$

s.t.

$$
\begin{aligned}
\tfrac{7}{10}x_1 + 1x_2 + 1s_1 &&&&&= 630 \\
\tfrac{1}{2}x_1 + \tfrac{5}{6}x_2 &&+\ 1s_2 &&&= 600 \\
1x_1 + \tfrac{2}{3}x_2 &&&&+\ 1s_3 &= 708 \\
\tfrac{1}{10}x_1 + \tfrac{1}{4}x_2 &&&&&+\ 1s_4 = 135
\end{aligned}
$$

$$x_1, x_2, s_1, s_2, s_3, s_4 \geq 0$$

We have used the s notation for the slack variables to make it easy to recall that these added variables are slack variables and not the original decision variables (x_1 and x_2). While the four slack variables could have been called x_3, x_4, x_5, and x_6, we feel the s notation is more descriptive.

Note also that we have expanded the nonnegativity constraints to include the four new slack variables. To see why the slack variables must also be constrained to nonnegative values, recall that our original cutting and dyeing constraint was

$$\tfrac{7}{10}x_1 + 1x_2 \leq 630. \tag{1.28}$$

Written in its equality form we had

$$\tfrac{7}{10}x_1 + 1x_2 + 1s_1 = 630. \tag{1.29}$$

What would it mean if s_1 were negative? A negative s_1 would mean that in order to satisfy equation (1.29), the quantity $\tfrac{7}{10}x_1 + 1x_2$ would have to be greater than 630. Since this would violate the original cutting and dyeing constraint, (1.28), the negative s_1 values are unacceptable. In addition, negative slack variables would indicate a negative excess capacity which has no physical interpretation. Thus, the nonnegativity constraints apply to all slack variables as well as the original decision variables.

As a final point, it is important for you to realize the Standard form of a linear programming problem is equivalent to the original formulation of the problem. That is, the optimal solution to any linear programming problem is the same as the optimal solution to the Standard form of the problem. Thus Standard form has not changed our basic problem; it has only changed how we write the constraints for the problem. Writing a linear programming problem in its Standard form is a necessary step to obtain the set of linear equations required by the Simplex solution procedure.

1.9   A General Notation for the Linear Programming Problem

Up to now we have confined our discussion of linear programming models solely to the Par, Inc. problem. We will now extend the Par, Inc. example in order to show you how to write the general form for any linear programming problem. You will find that the knowledge gained in learning how to write the general form will be applied in Chapter 2 when we show how the Simplex method can be used to solve linear programming problems. In addition, we will make extensive use of the general notation developed in this section in order to discuss the topics of duality and sensitivity analysis as presented in Chapter 4, the general transportation and assignment models as presented in Chapter 5, and the large-scale application of Chapter 6.

Generalizing the Par, Inc. Example

In Section 1.4, we formulated the following mathematical model of the Par, Inc. problem.

$$\max\ 10x_1 + 9x_2$$

s.t.

$$\tfrac{7}{10}x_1 + 1x_2 \le 630 \quad \text{Cutting \& Dyeing}$$
$$\tfrac{1}{2}x_1 + \tfrac{5}{6}x_2 \le 600 \quad \text{Sewing}$$
$$1x_1 + \tfrac{2}{3}x_2 \le 708 \quad \text{Finishing}$$
$$\tfrac{1}{10}x_1 + \tfrac{1}{4}x_2 \le 135 \quad \text{Inspection \& Packaging}$$
$$x_1,\ x_2 \ge 0$$

This model can only be used to represent one very special linear programming problem: the Par, Inc. problem. However, by introducing the following symbols to take the place of the coefficients in the Par, Inc. problem, we can write the general form for any linear programming problem consisting of two variables and four constraints.

c_j = objective function coefficient for the jth decision variable; $j = 1, 2$

b_i = right-hand side value for the ith constraint; $i = 1, 2, 3, 4$

a_{ij} = left-hand side value associated with constraint i and decision variable j; $i = 1, 2, 3, 4$; $j = 1, 2$

With the introduction of this notation, the general two-variable, four-constraint linear programming problem can be written as follows:

max $c_1x_1 + c_2x_2$

s.t.

$$a_{11}x_1 + a_{12}x_2 \leq b_1$$
$$a_{21}x_1 + a_{22}x_2 \leq b_2$$
$$a_{31}x_1 + a_{32}x_2 \leq b_3$$
$$a_{41}x_1 + a_{42}x_2 \leq b_4$$
$$x_1, \ x_2 \geq 0.$$

Our Par, Inc. problem is just one of many linear programs that can be written in this form. That is, the Par, Inc. problem is just the special case where

$$b_1 = 630 \quad a_{11} = 7\!/\!10 \quad a_{12} = 1$$
$$c_1 = 10 \quad b_2 = 600 \quad a_{21} = \frac{1}{2} \quad a_{22} = 5\!/\!6$$
$$c_2 = 9 \quad b_3 = 708 \quad a_{31} = 1 \quad a_{32} = \frac{2}{3}$$
$$b_4 = 135 \quad a_{41} = 1\!/\!10 \quad a_{42} = \frac{1}{4}.$$

To extend our general model for linear programs with 2 variables and 4 constraints to one which can be used to represent a

linear program consisting of any number of variables and constraints, we let

> m = the number of constraints in the linear programming
>   model, and

> n = the number of variables in the linear programming
>   model.

Hence, we can now write the general form of a linear programming problem involving m less-than-or-equal-to constraints and
n decision variables.

$$\max \ c_1x_1 + c_2x_2 + \cdots + c_nx_n \qquad \text{Objective Function}$$

s.t.

$$a_{11}x_1 + a_{12}x_2 + \cdots + a_{1n}x_n \le b_1 \qquad \text{Constraint 1}$$
$$a_{21}x_1 + a_{22}x_2 + \cdots + a_{2n}x_n \le b_2 \qquad \text{Constraint 2}$$

$$\vdots$$

$$a_{m1}x_1 + a_{m2}x_2 + \cdots + a_{mn}x_n \le b_m \qquad \text{Constraint m}$$
$$x_1, x_2, \ldots, x_n \ge 0 \qquad \text{Nonnegativity Restrictions}$$

In the above formulation the notation $(+ \ldots +)$ indicates
that the summation includes all of the terms between 2 and n.
Note that the Par, Inc. problem is just a special case of this
general model with m = 4, n = 2, and the values of the coefficients as given previously.

In the remaining part of this section we will show you how a
more concise form of the general linear programming problem
can be written in terms of matrix notation. That is, we will
show that an equivalent way of writing our general form is

$$\max \ c^t x$$

s.t.

$$Ax \le b$$
$$x \ge 0.$$

Before developing this alternate representation of the general linear programming problem, we will briefly review the necessary matrix notation and matrix operations.

Matrix Notation

We define a matrix to be a rectangular arrangement of numbers. For example, the following arrangement of numbers is a matrix named D.

$$D = \begin{bmatrix} 1 & 3 & 2 \\ 0 & 4 & 5 \end{bmatrix}$$

The matrix D is said to consist of six elements, where each element of D is a number. In order to identify a particular element of a matrix, we will have to specify its precise location. To do this, we introduce the notions of a row and a column.

All elements across some horizontal line in a matrix are said to be in a row of the matrix. For example, the elements 1, 3, and 2 in the matrix D are in the first row of D, and the elements 0, 4, and 5 are in the second row of D. Thus, we see that D is a matrix that has 2 rows. By convention, we always refer to the top row as row 1, the 2nd row from the top as row 2, etc.

All elements along some vertical line are said to belong to a column of the matrix. The elements 1 and 0 in the matrix D are elements in the first column of D, the elements 3 and 4 are elements of the second column, and the elements 2 and 5 are elements of the third column. Thus, we see that the matrix D has 3 columns. By convention, we always refer to the leftmost column as column 1, the next column to the right of column 1 as column 2, etc.

Thus, an easy way of identifying a particular element in a matrix is to specify its row and column position. For example, the element in row 1 and column 2 of the matrix D is the number 3. This is written as

$$d_{12} = 3.$$

In general, we use the following notation to refer to specific elements of matrices:

d_{ij} = element located in the ith row and jth column of D.

We will always use capital letters for the names of matrices and lower-case script versions of the same letter with two subscripts to denote the elements.

The *size* of a matrix is defined to be the number of rows and columns in the matrix and is written as "the number of rows x the number of columns". Thus, the size of the matrix D above is 2 x 3.

Frequently, we will encounter matrices that have only one row or one column. For example,

$$G = \begin{bmatrix} 6 \\ 4 \\ 2 \\ 3 \end{bmatrix}$$

is a matrix that has only one column. Whenever we have a matrix that has only one column like G above, we call the matrix a column vector. In a similar manner, any matrix that has only one row is called a row vector. Using our previous notation for elements of a matrix, we could refer to specific elements in G by writing g_{ij}. However, since G has only one column, the column position is unimportant, and we only need specify the row the element of interest is in. That is, instead of referring to elements in a vector using g_{ij}, we only specify one subscript which denotes the position of the element in the vector. For example,

$$g_1 = 6$$
$$g_2 = 4$$
$$g_3 = 2$$
$$g_4 = 3.$$

From now on, we will use a single subscript to identify specific elements in a vector, whether it be a column or a row vector. In addition, to clearly identify that we are referring to a vector, we will always name the vector with a lower-case letter. Thus, instead of referring to our above vector as G, we will use the lower-case letter g. In general, unless otherwise stated we will assume a vector is always written as a column vector.

In summary, an upper-case letter will denote a matrix, and a lower-case letter without a subscript will denote a vector. A lower-case, script letter with two subscripts will denote an element of a matrix, and a lower-case, script letter with one subscript will denote an element of a vector.

Matrix Operations

(1) Matrix Transpose—Given any matrix, we can form the transpose of the matrix by making the rows in the original matrix the columns in the transpose matrix, and by making the columns in the original matrix the rows in the transpose matrix. For example, if we take the transpose of the matrix

$$D = \begin{bmatrix} 1 & 3 & 2 \\ 0 & 4 & 5 \end{bmatrix}$$

we get

$$D^t = \begin{bmatrix} 1 & 0 \\ 3 & 4 \\ 2 & 5 \end{bmatrix}$$

Here you will note that we use the superscript t to denote the transpose of a matrix.

(2) Matrix Multiplication—We must learn how to perform two types of matrix multiplication. First, we must learn how to multiply two vectors, and then how to multiply a matrix times a vector.

The product of a row vector of size $1 \times n$ times a column vector of size $n \times 1$ is the number obtained by multiplying the first element in the row vector times the first element in the column vector, the second element in the row vector times the second element in the column vector, and continuing on through the last element in the row vector times the last element in the column vector, and then summing the products. Suppose, for example, that we wanted to multiply the row vector h times the column vector g where

$$h = [2 \quad 1 \quad 5 \quad 0] \text{ and } g = \begin{bmatrix} 6 \\ 4 \\ 2 \\ 3 \end{bmatrix}$$

The product gh is given by

$$gh = 2(6) + 1(4) + 5(2) + 0(3) = 26.$$

The product of a matrix of size m x n and a vector of size n x 1 is a new vector of size m x 1. The element in the ith position of the new vector is given by the vector product of the ith row of the m x n matrix times the m x 1 column vector. Suppose, for example, we want to multiply D times k where

$$D = \begin{bmatrix} 1 & 3 & 2 \\ 0 & 4 & 5 \end{bmatrix} \quad \text{and } k = \begin{bmatrix} 1 \\ 4 \\ 2 \end{bmatrix}$$

The first element of Dk is given by the vector product of the first row of D times k. Thus, we get

$$[1 \quad 3 \quad 2] \begin{bmatrix} 1 \\ 4 \\ 2 \end{bmatrix} = 1(1) + 3(4) + 2(2) = 17.$$

The second element of Dk is given by the vector product of the second row of D and k. Thus, we get

$$[0 \quad 4 \quad 5] \begin{bmatrix} 1 \\ 4 \\ 2 \end{bmatrix} = 0(1) + 4(4) + 5(2) = 26.$$

Hence we see that the product of the matrix D times the vector k is given by

$$Dk = \begin{bmatrix} 1 & 3 & 2 \\ 0 & 4 & 5 \end{bmatrix} \begin{bmatrix} 1 \\ 4 \\ 2 \end{bmatrix} = \begin{bmatrix} 17 \\ 26 \end{bmatrix}$$

Can any matrix and vector be multiplied? The answer is obviously, no! For us to multiply a matrix times a vector, the number of columns in the matrix must equal the number of rows in the vector. If this property is satisfied, the matrix and the vector are said to conform for multiplication. Thus, in our example, D and k could be multiplied because D had 3 columns and k had 3 rows.

Matrix Representation of the General Linear Programming Problem

Given the matrix notation and matrix operations we have just defined, let us now write the general linear programming problem in matrix notation. To do so, we define the following matrix A, and vectors b, x and c^t.

$$A = \begin{bmatrix} a_{11} & a_{12} & \cdot & \cdot & \cdot & a_{1n} \\ a_{21} & a_{22} & \cdot & \cdot & \cdot & a_{2n} \\ \cdot & \cdot & \cdot & \cdot & \cdot & \cdot \\ \cdot & \cdot & \cdot & \cdot & \cdot & \cdot \\ \cdot & \cdot & \cdot & \cdot & \cdot & \cdot \\ a_{m1} & a_{m2} & & & & a_{mn} \end{bmatrix}$$

$$b = \begin{bmatrix} b_1 \\ b_2 \\ \cdot \\ \cdot \\ \cdot \\ b_m \end{bmatrix}$$

$$x = \begin{bmatrix} x_1 \\ x_2 \\ \cdot \\ \cdot \\ \cdot \\ x_n \end{bmatrix}$$

$$c^t = \begin{bmatrix} c_1 & c_2 & \cdots & c_n \end{bmatrix}$$

The elements of these vectors and this matrix are given the same interpretation here that they were in our earlier formulation of the general linear programming problem. That is,

c_j = objective function coefficient for decision variable j; $j = 1, 2, \ldots, n$,

b_i = right-hand side value for constraint i; $i = 1, 2, \ldots, m$,

x_j = decision variable j; $j = 1, 2, \ldots, n$,

a_{ij} = coefficient of variable j in constraint i; $i = 1, 2, \ldots, m$; $j = 1, 2, \ldots, n$.

Thus, our general linear programming problem in matrix notation can be written as

max $c^t x$
s.t.

$Ax \leq b$
$x \geq 0$.

In the above representation $Ax \leq b$ means that every element of the vector Ax is less than or equal to the corresponding element of the vector b. Similarly, $x \geq 0$ means that every element of the vector x is greater than or equal to zero.

The above matrix representation is just a shorthand way of writing our general linear programming model. That is, if we actually carried out the vector and matrix vector multiplication indicated, we would get the general form representation that appeared at the beginning of this section. We suggest that the reader actually carry out the indicated matrix operations to verify the equivalence. Hence, we see that matrix representation provides us with a concise way of referring to the general linear programming model.

Note that we could extend c, x, and A such that c would include the objective function coefficients for the m slack variables, x would include the m slack variables, and A would include the m unit columns (i.e. columns containing a "1" as the only nonzero element)' corresponding to the m slack variables. If this were done, the Standard form of our general linear programming problem could be written as

max $c^t x$
s.t.

$$Ax = b$$
$$x \geq 0.$$

Making the above extensions for the Standard form of the Par, Inc. problem we would get the following c vector, b vector, and A matrix.

$$c^t = \begin{bmatrix} 10 & 9 & 0 & 0 & 0 & 0 \end{bmatrix} \qquad b = \begin{bmatrix} 630 \\ 600 \\ 708 \\ 135 \end{bmatrix}$$

$$A = \begin{bmatrix} 7/10 & 1 & 1 & 0 & 0 & 0 \\ 1/2 & 5/6 & 0 & 1 & 0 & 0 \\ 1 & 2/3 & 0 & 0 & 1 & 0 \\ 1/10 & 1/4 & 0 & 0 & 0 & 1 \end{bmatrix}$$

In the remainder of the text we shall often find it convenient to refer to the c vector, the b vector, and/or A matrix when describing a linear program. Whenever this is done, we shall always be thinking of the c vector as containing the objective function coefficients, the b vector as containing the values for the right-hand sides of the constraints, and the A matrix as containing the coefficients of the variables on the left-hand side of the constraints.

1.10 Summary

We have seen how one particular problem, the Par, Inc. golf bag production problem, could be formulated as a linear program and solved by a graphical procedure. In studying the graphical solution procedure we noted that the optimal solution to every linear programming problem occurs at one of the extreme points of the feasible region.

In the process of formulating a mathematical model of the Par, Inc. problem we developed the following general definition of a linear program:

> A linear program is a mathematical model which has the following properties:
>
> 1. a linear objective function which is to be maximized or minimized,
> 2. a set of linear constraints,
>
> and
>
> 3. variables which are all restricted to non-negative values.

We pointed out that the assumptions of proportionality and additivity must be satisfied by the real-world problem in order for the corresponding mathematical model to consist of a linear objective function and linear constraints. Thus it is only when these assumptions are satisfied that linear programming models are appropriate.

Although the algebraic solution to linear programs using the Simplex method will not be presented in detail until Chapter 2, it was pointed out that simultaneous linear equations will be an important part of this solution procedure. We saw how elementary row operations could be used to find solutions to such systems of linear equations. In addition, we learned that the linear equations encountered when applying the Simplex method will have an infinite number of solutions when the number of variables exceeds the number of constraint equations and a unique solution when the number of variables is the same as the number of equations.

The concept of a slack variable was introduced to show how less-than-or-equal-to constraints could be written in equality form. As we saw in section 1.8, when all of the constraints have been written as equalities, the linear program has been written in its Standard form.

Matrix representation of linear programs was presented because of its common usage and because it will be helpful in developing new linear programming concepts in the following chapters. The matrix representation of a linear program in Standard form is given by:

$$\max c^t x$$
s.t.
$$Ax = b$$
$$x \geq 0.$$

1.11   Glossary

1. *Objective function*—*All* linear programs have a linear objective function that is either to be maximized or minimized. In most linear programming problems the objective function will be used to measure the profit or cost of a particular solution.

2. *Constraint*—An equation or inequality which rules out certain combinations of variables as feasible solutions.

3. *Constraint function*—The left-hand side of a constraint relationship (i. e., the portion of the constraint containing the variables).

4. *Nonnegativity constraints*—A set of constraints that requires all variables to be nonnegative.

5. *Solution*—Any set of values for the variables.

6. *Feasible solution*—A solution which satisfies all of the constraints.

7. *Feasible region*—The set of all possible feasible solutions.

8. *Optimal solution*—A feasible solution that maximizes or minimizes the value of the objective function.

9. *Redundant constraint*—A constraint which does not affect the feasible region. If a constraint is redundant it could be removed from the problem without affecting the feasible region.

10. *Proportionality assumption*—The profit, cost, time, etc., associated with a variable is proportional to the value of the variable.

11. *Additivity assumption*—The total profit, total cost, total time, etc., of several variables can be found by adding the profit, cost, time, etc., associated with each of the variables.

12. *Linear equations or functions*—Mathematical expressions that satisfy the proportionality and additivity assumptions. Such expressions will have the variables in separate terms and in the first power.

13. *Mathematical model*—A representation of a problem where the objective and all constraint conditions are described by mathematical expressions.

14. *Linear program*—A mathematical model with a linear objective function, a set of linear constraints, and nonnegative variables.

15. *Extreme point*—Graphically speaking, extreme points are the feasible solution points occurring at the vertices or "corners" of the feasible region. With two variables, extreme points are determined by the intersection of the constraint lines. An important property of extreme points is that the optimal solution of a linear program always occurs at an extreme point.

16. *Simultaneous linear equations*—A system of linear equations which has a solution only if a set of values for the variables satisfies all equations.

17. *Elementary row operations*—A procedure for finding solutions to a system of simultaneous linear equations.

18. *Standard form*—A linear program in which all of the constraints are written as equalities. The optimal solution of the Standard form of a linear program is the same as the optimal solution of the original formulation of the linear program.

19. *Slack variable*—A variable added to the left-hand side of a less-than-or-equal-to constraint to convert the constraint into an equality. The value of this variable can usually be interpreted as the amount of unused resource.

20. *Matrix*—A rectangular arrangement of numbers.

21. *Vector*—A row vector is a matrix with only one row. A column vector is a matrix with only one column.

22. *Transpose*—The transpose of a matrix A, denoted A^t, is the matrix formed by interchanging the rows and columns of A.

23. *Size*—The size of a matrix is defined to be the number of rows and columns in the matrix and is written as "the number of rows x the number of columns". For example, if a matrix has 4 rows and 2 columns, the size of the matrix is written as 4 x 2.

24. *Conform*—A matrix A and a vector x are said to conform for multiplication if the number of columns in the matrix A is equal to the number of rows in the vector x. For example, if the size of A is 4 x 2 and the size of x is 2 x 1, Ax can be calculated since A and x conform for multiplication. A similar definition holds for two vectors.

25. *Vector multiplication*—If two vectors conform for multiplication, the product is the number obtained by multiplying and summing corresponding elements.

26. *Matrix vector multiplication*—The product of a matrix of size m x n and a vector of size n x 1 is a new vector of size m x 1. The element in the ith position of the new vector is given by the vector product of the ith row of the m x n matrix times the n x 1 column vector.

27. *Matrix representation of a linear program*—Writing a linear program in terms of vectors and matrices such as

max $c^t x$

s.t.

$$Ax = b$$
$$x \geq 0.$$

1.12 Problems

1. Consider the following linear programming problem.

max $1x_1 + 1x_2$

s.t.

$$1x_1 + 2x_2 \leq 6$$
$$6x_1 + 4x_2 \leq 24$$
$$x_1, x_2 \geq 0.$$

a. Find the optimal solution using the graphical procedure.

b. If the objective function were changed to $1x_1 + 3x_2$, what would be the optimal solution?

c. How many extreme points are there on your graph? What are the x_1 and x_2 values at each extreme point?

2. Consider the following linear programming problem.

max $2x_1 + 3x_2$

s.t.

$$3x_1 + 3x_2 \leq 12$$
$$\tfrac{2}{3}x_1 + x_2 \leq 4$$
$$x_1 + 2x_2 \leq 6$$
$$x_1, x_2 \geq 0.$$

a. Find the optimal solution using the graphical procedure.

b. Does this problem have a redundant constraint? If so, what is it? Does the solution change if the redundant constraint is removed from the problem?

3. Suppose that the management of Par, Inc. encounters each of the following situations:

    a. The accounting department revises its profit estimate on the Deluxe bags to $18 per bag.

    b. A new low-cost material is available for the Standard bag, and the profit per Standard bag can be increased to $20 per bag. (Assume the profit of the Deluxe bag is the original $9 value.)

    c. New sewing equipment is available and would increase the sewing operation capacity to 750 hours. (Assume $10x_1 + 9x_2$ is the appropriate objective function.)

    If each of the above conditions is encountered separately, what do you recommend under each case? That is, what is the optimal solution and profit for each situation?

4. For the following equations and inequalities

    1. $6x_1 - 2x_2 + \tfrac{1}{2}x_3 \leq 100$

    2. $-1x_1 + 2x_2{}^2 + 1x_3 = 50$

    3. $2x_1 \quad\quad + 5x_3 \leq 10$

    4. $1x_1 + 2x_2 + 1x_1x_2 = 25$

    5. $3\sqrt{x_1} + 2x_2 + 1x_3 \leq 5$

    6. $1x_1 + 1x_2 + 1x_3 \leq 20$

    a. Which of the above mathematical relationships could be found in a linear programming model and which could not? For the relationships that are unacceptable for linear programs, state your reasons.

    b. What are the two assumptions that must be met by the objective function and constraints of every linear program?

5. Suppose that the time it takes to assemble one Sharpshooter rifle is 1 minute but that 150 of these rifles can be assembled in an hour.

    a. Is the assembly time for these rifles proportional to the number of rifles produced? Why or why not?

    b. Would a linear programming model which included a constraint on the amount of assembly time available be valid?

6. Grippo Golf Glove Company makes two different brands
    of golf gloves. One is a full-fingered glove and the other
    is a half-fingered model. Grippo currently has orders for
    more gloves than it can produce in time for the upcoming
    golf season. The scarce resource in the manufacture of
    these gloves is labor time.

    Grippo has available 400 hours in the cutting and sew-
    ing department, 250 hours in the finishing department, and
    150 hours in the packaging and shipping department.

    The department time requirements and the profit per
    box (1 gross) are given below.

	Cutting & Sewing	Finish- ing	Packaging & Shipping	Profit
full-finger	3 hrs.	1½ hrs.	1 hr.	$30
half-finger	1½ hrs.	2 hrs.	1 hr.	$25

Find the optimal product mix for Grippo assuming the company wants to maximize profit.

What is the maximum amount of profit Grippo can earn?

7. The URNUTZ Company manufactures and sells two prod-
    ducts. The company makes a profit of $7 for each unit of
    product 1 and a profit of $6 for each unit of product 2 sold.
    The man-hours requirements for the products in each of the
    three production departments are summarized below:

	Product 1	Product 2
Dept. 1	⅓ hr.	½ hr.
Dept. 2	¼ hr.	⅙ hr.
Dept. 3	½ hr.	⅜ hr.

The supervisors of these departments have estimated that
the following number of man-hours will be available during the next month: 200 man-hours in department 1, 150
man-hours in department 2, and 500 man-hours in department 3.

Assuming that URNUTZ is interested in maximizing profits,

a. show the linear programming model of this problem;

b. find the optimal solution using the graphical procedure.

8. The Sweet Tooth Candy Company has a limited supply of two chocolate ingredients that are used in the production of candy bar products. There are 8000 ounces available of XJ100 and 7000 ounces available of XJ200. Both ingredients are made especially for Sweet Tooth by the Marcus T. Cavity Chocolate Compound Company. Sweet Tooth uses the two ingredients to produce either Crunch-A-Munch Bars or Golden-Goodie Bars. Each Crunch-A-Munch Bar uses $\frac{1}{4}$ of an ounce of XJ100 and $\frac{1}{2}$ of an ounce of XJ200. Each Golden-Goodie Bar requires $\frac{1}{2}$ of an ounce of XJ100 and $\frac{1}{4}$ of an ounce of XJ200. In addition, the marketing department of Sweet Tooth estimates they will be able to sell at most 10,000 Crunch-A-Munch Bars. If Sweet Tooth makes a profit of $0.05 on each Crunch-A-Munch and $0.02 on each Golden Goodie, what product mix should the company employ? Solve this problem using the graphical procedure.

9. Use elementary row operations to find the solution to the following set of linear equations:

$$1x_1 - 2x_2 + 3x_3 = 9$$
$$2x_1 - x_2 + x_3 = 4$$
$$x_1 + 2x_2 - x_3 = 1.$$

10. Use elementary row operations to find solutions to the following set of linear equations:

$$2x_1 + 4x_2 - 2x_3 = 20$$
$$x_1 - x_2 + 4x_3 = 5.$$

Show at least two specific solutions for this set of linear equations.

11. Suppose you were given the following system of linear equations

$$2x_1 + x_2 + 3x_3 = 5$$
$$5x_1 - x_3 = 7.$$

a. How many solutions are there to this system of equations?

b. Suppose x_3 is set equal to zero. How many solutions are there now? Find all of these solutions.

c. Suppose x_2 is set equal to zero. How many solutions are there now? Find all of these solutions.

12. What constraint lines combine to form extreme point ④ of the Par, Inc. problem (see Figure 1–N)? Use the solution of simultaneous linear equations to show that the exact values of x_1 and x_2 at this extreme point are $x_1 = 300$ and $x_2 = 420$.

13. An oil refinery in Tulsa, Oklahoma sells 60% of its oil production to a distributor in Chicago and 40% to a distributor in Atlanta. Another refinery in New Orleans, Louisiana sells 30% of its oil production to the same Chicago distributor and 70% to the same Atlanta distributor. If we know the Chicago distributor received 120,000 gallons of oil from the two plants and the Atlanta distributor received 180,000 gallons from the two plants during the last month, how many gallons of oil were produced at each of the two plants? (Hint: Let $x_1 =$ gallons at the Tulsa plant and $x_2 =$ gallons at the New Orleans plant. Then define a linear equation for each distributor and solve the two simultaneous linear equations.)

14. The Death-to-Weeds Company is a major manufacturer of weed killers for home use. They currently have available 175 pounds of K–20 and 900 pounds of K–25 weed killing concentrates used in the manufacture of a number of Death-to-Weeds products. The company will use these concentrates in the production of Dandelion-Do-In and Quack-Killer, two of the most profitable company products. Each unit of Dandelion-Do-In uses ¼ of a pound of K–20 and 1 pound of K–25. Each unit of Quack-Killer requires ⅙ of a pound of K–20 and 1½ pounds of K–25. The company has enough packaging material to produce as much of Dandelion-Do-In as desired but is limited to a maximum of 400 units of Quack-Killer. If the profit contribution for both products is $6 per unit produced, how many units of each product should the company manufacture? Solve this problem using the graphical procedure.

15. Write the following linear program in Standard form.

   max $5x_1 + 2x_2 + 8x_3$

   s.t.

   $$1x_1 - 2x_2 + \tfrac{1}{2}x_3 \le 420$$
   $$2x_1 + 3x_2 - 1x_3 \le 610$$
   $$6x_1 - 1x_2 + 3x_3 \le 125$$
   $$x_1, x_2, x_3 \ge 0.$$

16. Given the following linear program

   max $3x_1 + 4x_2$

   s.t.

   $$-1x_1 + 2x_2 \le \ 8$$
   $$1x_1 + 2x_2 \le 12$$
   $$2x_1 + \ x_2 \le 16$$
   $$x_1, x_2 \ge 0.$$

   a. Write this problem in its Standard form.
   b. Solve the above problem using the graphical procedure.
   c. What are the values of the three slack variables at the optimal solution?

17. For the following linear program

   max $10x_1 + 2.5x_2$

   s.t.

   $$5x_1 + \quad x_2 \le 15$$
   $$6x_1 + \ 4x_2 \le 24$$
   $$x_1 + \quad x_2 \le \ 5$$
   $$x_1, x_2 \ge 0.$$

   a. Write this problem in its Standard form.
   b. Solve the above problem using the graphical procedure.
   c. What the the values of the three slack variables at the optimal solution?

18. Speen Food Supplies, Inc. is a manufacturer of frozen pizzas. Art Speen, president of Speen Food Supplies, personally supervises the production of both types of frozen pizzas produced by the company: Speen's Regular, and Speen's Super-Deluxe. Art makes a profit of $0.50 for each Regular produced and $0.75 for each Super-Deluxe. He currently has 150 pounds of dough mix available and 800 ounces of topping mix. Each Regular pizza uses 1 pound of dough mix and 4 ounces of topping, whereas each Super-Deluxe uses 1 pound of dough mix and 8 ounces of topping mix. Based upon past demand, Art knows that he can sell at most 75 Super-Deluxe pizzas and 125 Regular pizzas. How many Regular and Super-Deluxe pizzas should Art make in order to maximize profits?

   a. Solve this problem graphically using the Simplex method.

   b. Show the problem in its Standard form.

   c. What are the values and interpretations of all slack variables? Which constraints are binding the optimal solution?

19. Wilkinson Motors, Inc. sells standard automobiles and station wagons. The firm makes $200 profit for each automobile it sells and $250 profit for each station wagon it sells. The company is planning next month's order which the manufacturer says cannot exceed 300 automobiles and 150 station wagons. Dealer preparation time requires 2 hours for each automobile and 3 hours for each station wagon. Next month the company has 900 hours of shop time available for new car preparation. How many automobiles and station wagons should be ordered so that profit is maximized?

   a. Show the linear programming model of the above problem.

   b. Show the Standard form and identify the slack variables.

   c. Identify the extreme points of the feasible region.

   d. Solve graphically.

20. Ryland Farms in northwestern Indiana grows soybeans and
     corn on its 500 acres of land. An acre of soybeans brings
     a $50 profit and an acre of corn brings a $100 profit. Be-
     cause of government regulations, no more than 200 acres
     can be planted in soybeans. During the planting season
     1200 man-hours of planting time will be available. Each
     acre of soybeans requires 2 man-hours while each acre of
     corn requires 6 man-hours. How many acres of soybeans
     and how many acres of corn should be planted in order to
     maximize profits?

     a. Show the linear programming model of the above prob-
         lem.

     b. Show the Standard form and identify all slack variables.

     c. Solve graphically.

     d. Identify all the extreme points of the feasible region.

     e. If the farm could get either more man-hours of labor
         or additional land, which should it attempt to obtain?
         Why?

21. Let

$$A = \begin{bmatrix} 6 & 2 & 3 \\ 4 & 8 & 1 \end{bmatrix} \qquad c = \begin{bmatrix} 1 \\ 7 \\ 0 \end{bmatrix}$$

$$x = \begin{bmatrix} x_1 \\ x_2 \\ x_3 \end{bmatrix} \qquad b = \begin{bmatrix} 30 \\ 20 \end{bmatrix}$$

     a. Find Ax.

     b. Find cx.

     c. Write the linear program corresponding to the above
         A matrix and c, b, and x vectors.

22. Consider again the linear program appearing in problem 2:

$$\max \ 2x_1 + 3x_2$$

s.t.

$$3x_1 + 3x_2 \leq 12$$
$$\tfrac{2}{3}x_1 + \ x_2 \leq \ 4$$
$$x_1 + 2x_2 \leq \ 6$$
$$x_1, x_2 \geq 0.$$

Define the following:

a.   the A matrix associated with the current form of the problem,

b.   the c and b vectors for the current form of the problem,

c.   the Standard form for this problem, and

d.   the A matrix and the c and b vectors associated with the Standard form of this problem.

23. If

$$c = \begin{bmatrix} 10 \\ 8 \\ 12 \end{bmatrix} \qquad b = \begin{bmatrix} 100 \\ 200 \\ 250 \end{bmatrix}$$

and

$$A = \begin{bmatrix} \tfrac{1}{2} & 3 & -1 \\ 1 & 1 & 2 \\ 4 & 0 & 1 \end{bmatrix}$$

a.   Write the complete mathematical model of the linear program

$$\max \ c^t x$$

s.t.

$$Ax \leq b$$
$$x \geq 0.$$

b.   Show the new x, c and b vectors and the new A matrix
if we were to write this problem in Standard form

max $c^t x$

s.t.

$$Ax = b$$
$$x \geq 0.$$

24.   Show the Speen Food Supplies problem (18) as a matrix
representation of the initial formulation and the Standard
form.   Be sure to show x, c, b, and A for both cases.

25.   The marketing department of KT Company is interested
in finding out how to get the most audience exposure from
its current advertising budget.   The company would like
to determine how much of its advertising budget should
be spent on each of three media: radio, television, and newspaper.

Each dollar spent on radio advertising is worth 6 exposure points to the company.   Similarly, KT believes it will
get 5 and 8 exposure points respectively for every dollar
spent on television and newspaper advertising.

KT's total advertising budget consists of $10,000.   How-
ever, because of an agreement with a local television station, KT may not spend more than half as much on radio
advertising as it does on television.   In addition, the combined expenditure on television and newspaper advertising may not exceed 80% of the total advertising expenditure.

Assuming that KT is interested in maximizing its exposure points, formulate this problem as a linear program.
(Hint: Let $x_1 = $ dollars spent on radio, $x_2 = $ dollars spent
on television, and $x_3 = $ dollars spent on newspaper.)   Set
up the Standard form for this problem and interpret all
slack variables.   (You do not have to solve this problem.)

2

The Simplex Method

We have seen how to find the optimal solution for two-variable linear programming problems using the graphical procedure; however, most real-world problems contain more than two decision variables and are thus too large for this solution technique. An algebraic solution procedure—the Simplex method—will have to be used to solve these larger linear programming problems. Computer programs of the Simplex method have been used to solve linear programming problems having as many as several thousand variables and several thousand constraints. The computer program included in the Appendix is based upon the Simplex method and can be used to solve linear programming problems having up to 250 variables and 80 constraints.

In this Chapter we present the Simplex method in a step-by-step fashion using the Par, Inc. example of Chapter 1 to illustrate how the method works. After the method has been developed for this particular problem, we then set forth the general Simplex procedure which can be used to solve any linear program.

2.1   An Algebraic Overview of the Simplex Method

Let us return to the Par, Inc. problem which is written below in its Standard form.

$$\max 10x_1 + 9x_2 + 0s_1 + 0s_2 + 0s_3 + 0s_4 \tag{2.1}$$

s. t.

$$\begin{aligned}
\tfrac{7}{10}x_1 + 1x_2 + 1s_1 &&&= 630 && (2.2)\\
\tfrac{1}{2}x_1 + \tfrac{5}{6}x_2 &+ s_2 && = 600 && (2.3)\\
1x_1 + \tfrac{2}{3}x_2{\scriptstyle |} &&+ 1s_3 &= 708 && (2.4)\\
\tfrac{1}{10}x_1 + \tfrac{1}{4}x_2 &&&+ 1s_4 = 135 && (2.5)
\end{aligned}$$

$$x_1, x_2, s_1, s_2, s_3, s_4 \geq 0 \tag{2.6}$$

What is involved in finding the optimal solution to this problem algebraically?   First note that equations (2.2)–(2.5), the constraint equations, form a system of four simultaneous linear equations in six variables. In order to satisfy the constraints of the Par, Inc. problem, the optimal solution must also be a solution to this set of linear equations. Since there are an infinite number of solutions to this system of equations, we see that any algebraic procedure for solving linear programs must be capable of finding solutions to systems of simultaneous equations in which there are more variables than equations.

Second, note that not every solution to equations (2.2)–(2.5) is a feasible solution to the linear program. That is, we cannot expect every solution to equations (2.2)–(2.5) to also satisfy the nonnegativity conditions $(x_1, x_2, s_1, s_2, s_3, s_4 \geq 0)$. Thus we see that an algebraic procedure for solving linear programming problems should be capable of eliminating from consideration those solutions to equations $(2.2)-(2.5)$ which do not also satisfy the nonnegativity requirement.

Finally, an algebraic procedure for solving linear programs must be capable of picking one of these feasible solutions as the one which maximizes the objective function. The *Simplex method* is an algebraic procedure with all three of the capabilities outlined above.

We have stated that our system of equations $((2.2)$, (2.3), (2.4), and $(2.5))$ will have an infinite number of solutions. The way the Simplex method finds solutions for these equations is to assign zero values for any two of the variables and then solve for the values of the remaining four variables. To illustrate this procedure, suppose we set $x_2 = 0$ and $s_1 = 0$. Our system of equations then becomes

$$\frac{7}{10}x_1 \qquad\qquad\qquad\qquad = 630 \qquad\qquad (2.7)$$
$$\frac{1}{2}x_1 \quad + 1s_2 \qquad\qquad\quad = 600 \qquad\qquad (2.8)$$
$$1x_1 \qquad\quad + 1s_3 \qquad\quad = 708 \qquad\qquad (2.9)$$
$$\frac{1}{10}x_1 \qquad\qquad\quad + 1s_4 \; = 135. \qquad\qquad (2.10)$$

By setting $x_2 = 0$ and $s_1 = 0$ we have in effect reduced our system of linear equations to four variables and four equations. Using the elementary row operations introduced in Chapter 1 we can solve for the values of the remaining four variables.

Multiplying the first row by $\frac{10}{7}$ we have

$$1x_1 = 900.$$

Using additional row operations to obtain a zero coefficient for x_1 in all other equations, our system of linear equations reduces to

$$1x_1 \qquad\qquad\qquad\qquad = 900$$
$$1s_2 \qquad\qquad\qquad = 150$$
$$1s_3 \qquad\quad = -192$$
$$1s_4 \quad = \quad 45.$$

We now have one solution to our original set of four equations and six variables. This solution is

$$\begin{bmatrix} x_1 = & 900 \\ x_2 = & 0 \\ s_1 = & 0 \\ s_2 = & 150 \\ s_3 = & -192 \\ s_4 = & 45 \end{bmatrix}$$

Using the above procedure, we have found what is known as a *basic solution* to our linear programming problem. In general, if we have a Standard form linear programming problem consisting of n variables and m equations where n is greater than m, a *basic solution* can be obtained by setting n − m of the variables equal to zero and solving the m constraint equations for the remaining m variables. In terms of our Par, Inc. problem, a basic solution can be obtained by setting *any* two variables equal to zero and then solving the system of four equations for the remaining four variables. We shall refer to the n − m variables set equal to zero as the *nonbasic* variables and the remaining m variables (usually nonzero) as the *basic* variables. Thus, in the above example, x_2 and s_1 are the nonbasic variables and x_1, s_2, s_3 and s_4 are the basic variables.

Certainly, the above basic solution is not feasible since the nonnegativity conditions are not satisfied (i. e., s_3 is less than zero). Thus, we see that a basic solution does *not* have to be a feasible solution. However, when a basic solution is also feasible, we refer to it as a *basic feasible solution*. For example, if we set $x_1 = 0$ and $x_2 = 0$, we can solve for s_1, s_2, s_3, and s_4. Doing this, our system of equations becomes

$$
\begin{aligned}
s_1 & & & = 630 \\
& s_2 & & = 600 \\
& & s_3 & = 708 \\
& & & s_4 = 135.
\end{aligned}
$$

Thus, the solution corresponding to $x_1 = 0$ and $x_2 = 0$ is

$$
\begin{bmatrix}
x_1 = & 0 \\
x_2 = & 0 \\
s_1 = & 630 \\
s_2 = & 600 \\
s_3 = & 708 \\
s_4 = & 135
\end{bmatrix}
$$

Clearly, this solution represents a basic solution to our problem since it was obtained by setting two of the variables equal to zero and solving for the other four. Moreover, this solution is a basic feasible solution since each of the variables is greater than or equal to zero. Referring to Figure 2–A, we see that this basic feasible solution corresponds to extreme point 1 of the feasible region ($x_1 = 0$ and $x_2 = 0$). Thus, in this case, a basic feasible solution corresponds to an extreme point. This is not just a coincidence, but an important property of all basic feasible solutions. That is, basic feasible solutions always occur at the extreme points of the feasible region. *In other words, a basic feasible solution and an extreme point solution are one and the same!*

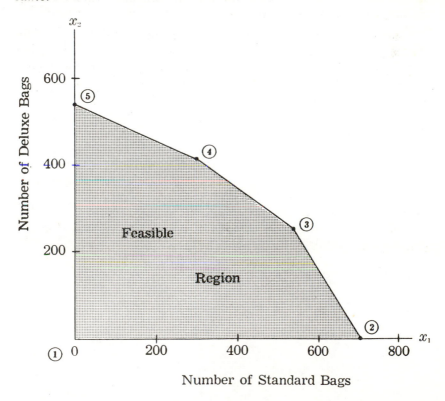

Figure 2–A. The Five Extreme Points of the Feasible Region for the Par, Inc. Problem

A basic feasible solution to the system of m constraint equations and n variables is required as a starting point for the Simplex method. Such a solution can be easily found by setting all the decision variables equal to zero. This corresponds to selecting the origin (extreme point 1 in the Par, Inc. problem) as the initial basic feasible solution for the Simplex procedure. From this starting point, the Simplex method successively generates basic feasible solutions to our system of equations, making sure that the objective function increases for each new solution. Since, as we saw in Chapter 1, the optimal solution to a linear programming problem always occurs at an extreme point, and since a basic feasible solution and an extreme point solution are synonomous, the Simplex method must eventually locate the optimal solution to the problem. Thus the Simplex method can be described as an iterative procedure for moving from one basic feasible solution (extreme point) to another until the optimal solution is reached. The way in which this iterative procedure is carried out is the subject of the remainder of this Chapter.

2.2   Tableau Form

As we discussed in the previous section, the Simplex method always begins with a basic feasible solution and then moves from one basic feasible solution to another until the optimal basic feasible solution (extreme point) is reached. Thus, prior to beginning the Simplex method, we must find an initial basic feasible solution for our system of constraint equations. Recall that for our Par, Inc. problem, the Standard form was

$$\max \quad 10x_1 + 9x_2 + 0s_1 + 0s_2 + 0s_3 + 0s_4$$

s.t.

$$
\begin{aligned}
\tfrac{7}{10}x_1 + 1x_2 + 1s_1 &&&& = 630 \\
\tfrac{1}{2}x_1 + \tfrac{5}{6}x_2 &+ 1s_2 &&& = 600 \\
1x_1 + \tfrac{2}{3}x_2 && + 1s_3 && = 708 \\
\tfrac{1}{10}x_1 + \tfrac{1}{4}x_2 &&& + 1s_4 & = 135 \\
\end{aligned}
$$

$$x_1\, x_2,\, s_1,\, s_2,\, s_3,\, s_4 \geq 0.$$

As we saw in the previous section, it is very easy to find an initial basic feasible solution for this problem. We simply set $x_1 = 0$ and $x_2 = 0$ and solve for s_1, s_2, s_3, and s_4. Thus, we obtain the solution $x_1 = 0$, $x_2 = 0$, $s_1 = 630$, $s_2 = 600$, $s_3 = 708$, and $s_4 = 135$. The reason this basic feasible solution was so easy to find is that as soon as x_1 and x_2 had been set equal to zero the values for the remaining variables could simply be read from the right-hand side of the constraint equations. If we study this particular system of equations closely, we can identify two properties that make it possible to easily find a basic feasible solution.

The first property enables us to easily find a basic solution. Loosely stated, this property says that m of the variables (m = 4 in this case) must have both a coefficient of one and appear with a nonzero coefficient in only one equation each. Then if these m variables are made basic by setting the other (n − m) variables equal to zero, the values of the basic variables can be read from the right-hand side of the constraint equations. In the example the variables s_1, s_2, s_3, and s_4 satisfy this first property. The second property enables us to easily find a basic *feasible* solution for a linear program. This property requires that the values on the right-hand side of the constraint equations be nonnegative. In our example, we see that this second property is also satisfied.

If we can write our linear programming problem in a form which satisfies the first property, then the values of the basic variables are given by the right-hand sides of the constraining equations. If, in addition, the second property is satisfied, the values of the variables will be nonnegative and the basic solution will also be feasible.

If a linear programming problem satisfies both of the above properties, it is said to be in *Tableau form*. We note that the Standard form representation of the Par, Inc. problem is already in Tableau form. In fact, the Standard form and Tableau form representations of linear programs that have all less-than-or-equal-to constraints and nonnegative right-hand side values are the same. However, as we shall see in the next chapter, there are a great many linear programming problems for which Standard form and Tableau form are not the same.

Let us pause for a moment and reflect on the reason for introducing the notion of Tableau form. Since the Simplex method always begins with a basic feasible solution and since the Tableau form provides an easy way of obtaining an initial basic feasible solution, putting a linear programming problem into Tableau form is an important step in preparing the problem for solution by the Simplex method. Thus, the following three steps are necessary in order to prepare a linear programming problem for solution using the Simplex method.

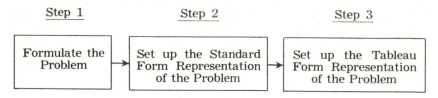

Step 1   Step 2   Step 3

| Formulate the Problem | Set up the Standard Form Representation of the Problem | Set up the Tableau Form Representation of the Problem |

2.3   Setting Up the Initial Simplex Tableau

After a linear programming problem has been converted to Tableau form, we have available an initial basic feasible solution which can be used to begin the Simplex method. The next step is to set up the initial *Simplex tableau.* The Simplex tableau provides a convenient means for keeping track of and performing the calculations necessary during the Simplex solution procedure.

Part of the initial Simplex tableau is simply a table containing all the information shown in the Tableau form representation of a linear program. This portion of the tableau is as follows:

c_1	c_2	. . . .	c_n	
a_{11}	a_{12}	. . . .	a_{1n}	b_1
a_{21}	a_{22}	. . . .	a_{2n}	b_2
.	.	. . . .	.	.
.	.	. . . .	.	.
.	.	. . . .	.	.
a_{m1}	a_{m2}	. . . .	a_{mn}	b_m

In the above partial tableau, the horizontal and vertical lines are used to separate the different parts of the Tableau form representation of a linear program. The upper horizontal line separates the coefficients of the variables in the objective function from the coefficients of the variables in the constraint equations. The vertical line can be interpreted as an equality line; the values to the left of this line are the coefficients of the variables in the constraint equations, and the values to the right of the line are the right-hand side values of the constraint equations.

Before we will be ready to apply the Simplex method, two more rows and two more columns will have to be added to our tableau. However, before defining these new rows and columns, let us set up the partial Simplex tableau for our Par, Inc. example problem. Tableau form (the same as Standard form in this case) for the Par, Inc. problem is

$$\max \quad 10x_1 + 9x_2 + 0s_1 + 0s_2 + 0s_3 + 0s_4$$

s.t.

$$\frac{7}{10}x_1 + 1x_2 + 1s_1 \qquad\qquad\qquad = 630$$
$$\frac{1}{2}x_1 + \frac{5}{6}x_2 \qquad + 1s_2 \qquad\qquad = 600$$
$$1x_1 + \frac{2}{3}x_2 \qquad\qquad + 1s_3 \qquad = 708$$
$$\frac{1}{10}x_1 + \frac{1}{4}x_2 \qquad\qquad\qquad + 1s_4 = 135$$
$$x_1, x_2, s_1, s_2, s_3, s_4 \geq 0 .$$

A partial Simplex tableau can then be written as

10	9	0	0	0	0	
7/10	1	1	0	0	0	630
1/2	5/6	0	1	0	0	600
1	2/3	0	0	1	0	708
1/10	1/4	0	0	0	1	135

Notice that the row above the first horizontal line contains the coefficients of the objective function for our Par, Inc. problem in Tableau form. The elements appearing between the horizontal lines and to the left of the vertical line are simply the co-efficients of the constraint equations; whereas the elements to the right of the vertical line are the right-hand side values of the constraint equations. It may be easier to recall that each of the rows contains the coefficients of one constraint equation if we note that each of the columns is associated with one of the variables. For example, x_1 corresponds to the first column, x_2 the

second, s_1 the third, and so on. To help us keep this in mind we
will write the variable associated with each column directly above
the column. Doing so, we get

x_1	x_2	s_1	s_2	s_3	s_4	
10	9	0	0	0	0	
7/10	1	1	0	0	0	630
1/2	5/6	0	1	0	0	600
1	2/3	0	0	1	0	708
1/10	1/4	0	0	0	1	135

Can you write the constraint equations contained in the partial
Simplex tableau of a different linear program? Try it!

x_1	x_2	s_1	s_2	
6	8	0	0	
2	5	1	0	11
4	1	0	1	14

Did you get the following?

$$2x_1 + 5x_2 + 1s_1 + 0s_2 = 11$$
$$4x_1 + 1x_2 + 0s_1 + 1s_2 = 14$$

We can also see that the objective function for this problem is
$6x_1 + 8x_2 + 0s_1 + 0s_2$.

We stated previously that the Simplex method must be started
with a basic feasible solution. Certainly, one basic feasible solution for the Par, Inc. problem is the one we found in the previous
section from the Tableau form of our problem. This solution

corresponds to a product combination of zero Standard bags and zero Deluxe bags and is represented by the solution vector

$$\begin{bmatrix} x_1 \\ x_2 \\ s_1 \\ s_2 \\ s_3 \\ s_4 \end{bmatrix} = \begin{bmatrix} 0 \\ 0 \\ 630 \\ 600 \\ 708 \\ 135 \end{bmatrix}$$

Note that the initial Simplex tableau contains the Tableau form of the problem and thus it is easy to find the above basic feasible solution from the initial Simplex tableau. As you can see, associated with each basic variable is a column in the Simplex tableau which has a "1" in the only nonzero position. Such columns are known as *unit columns* or *unit vectors*. Also associated with each basic variable is a row of the tableau. This row can be identified by the fact that it also contains the "1" in the unit column. The value of each basic variable is then given by b_i in the row associated with the basic variable. For example, s_3 has a "1" in row 3; therefore the value of this basic variable is given by $s_3 = b_3 = 708$. This procedure is shown in Table 2–A.

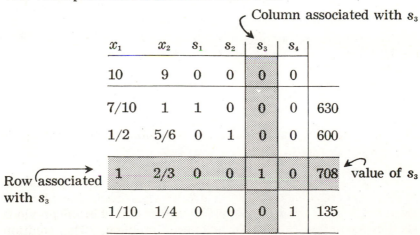

Table 2–A. Illustration of Procedure for Finding Values of Basic Variables from the Simplex Tableau.

At this point we have seen how one goes about finding an initial basic feasible solution and setting up a partial Simplex tableau. We know that the Simplex method proceeds from one basic feasible solution to another until the optimal basic feasible solution is reached. In the next section we discuss how the Simplex method moves from our initial basic feasible solution to a better one.

2.4 Improving the Solution

To improve the solution we will have to generate a new basic feasible solution (extreme point) that yields a higher profit. To do this we will have to change the set of basic variables. That is, we will have to select one of the current nonbasic variables to bring into solution and one of the current basic variables to leave the solution in such a fashion that the new basic feasible solution yields a higher value for the objective function. The Simplex method provides us with an easy way to carry out this change in the basic feasible solution.

For convenience, we will now add two new columns to the present form of the Simplex tableau in order to keep track of the basic variables and the profit associated with these variables. One column will be labeled BASIS, and the other labeled c_j. Under the column labeled BASIS, we shall list the names of the current basic variables, and under the column labeled c_j, we shall list the profit corresponding to each of these basic variables. For the Par, Inc. problem, this results in the following intial Simplex tableau.

BASIS	c_j	x_1	x_2	s_1	s_2	s_3	s_4	
		10	9	0	0	0	0	
s_1	0	7/10	1	1	0	0	0	630
s_2	0	1/2	5/6	0	1	0	0	600
s_3	0	1	2/3	0	0	1	0	708
s_4	0	1/10	1/4	0	0	0	1	135

We note in the BASIS column that s_1 is listed first since its value is given by b_1; s_2 second since its value is given by b_2; etc.

Can we improve upon our present basic feasible solution? To find out if this is possible, we introduce two new rows into our tableau. The first row, labeled z_j, will represent the *decrease* in the value of the objective function that will result if one unit of the variable corresponding to the jth column of the A matrix is brought into solution. For example, z_1 will represent the decrease in profit that will result if one unit of x_1 is brought into solution. Let us see why a decrease in profit might result if we bring x_1 into solution.

If one unit of x_1 is produced, we will have to change the value of some of the current basic variables in order to satisfy our constraint equations. In the first constraint equation we have

$$\frac{7}{10}x_1 + 1x_2 + 1s_3 = 630.$$

If we are considering making x_1 some positive value, we will have to reduce x_2 and/or s_3 in order to satisfy this constraint. Since x_2 is already zero (x_2 is a nonbasic variable), it cannot be reduced any further. Thus the value of s_3 will be reduced if x_1 is made positive. This reduction in the value of a basic variable may result in a reduction in the value of the objective function. The amount of the reduction depends of course upon the coefficient of s_3 in the objective function. In this case, since s_3 is a slack variable, its coefficient is zero; thus reducing s_3 will not lower the value of the objective function.

On the other hand, every unit of x_1 introduced will improve the value of the objective function by the amount c_1, which in our Par, Inc. problem is the $10 profit associated with each Standard bag produced. Since the value of the objective function will decrease by z_1 for each unit of x_1 produced, the net change in the value of the objective function that results due to one unit of x_1 being introduced is given by $c_1 - z_1$. The next row we introduce into our tableau, which we refer to as the *net evaluation row*, will contain the value of $c_j - z_j$ for every variable (column) in the tableau. In terms of position in the tableau, the z_j and $c_j - z_j$ rows will be placed directly under the A matrix in the existing tableau. Now, let us get back to the original question of which variable we should make basic by calculating the entries in the net evaluation row for the Par, Inc. problem.

If we were to bring one unit of x_1 into the solution, we see from analyzing the constraint equations that we would have to give up $\frac{7}{10}$ of an hour of cutting and dyeing time, $\frac{1}{2}$ of an hour of sewing time, 1 hour of finishing time, and $\frac{1}{10}$ of an hour of inspection and packaging time. Thus, we note that the coefficients in the x_1 column indicate how many units of the basic variable in that row will be driven out of solution when one unit of x_1 is brought in. In general, all of the column coefficients can be interpreted this way. Thus, if we were to bring one unit of x_2 into solution, we would have to give up one unit of s_1, $\frac{5}{6}$ of a unit of s_2, $\frac{2}{3}$ of a unit of s_3 and $\frac{1}{4}$ of a unit of s_4.

To calculate how much the objective function will decrease when one unit of a nonbasic variable is brought into solution, we must know the value of the objective function coefficients for the basic variables. These values are given in the c_j column of our tableau. Hence, *the values in the z_j row can be calculated by multiplying the elements in the c_j column by the corresponding elements in the columns of the A matrix and summing them.* Thus, we get

$$z_1 = 0(\tfrac{7}{10}) + 0(\tfrac{1}{2}) + 0(1) + 0(\tfrac{1}{10}) = 0$$
$$z_2 = 0(1) \;\;\; + 0(\tfrac{5}{6}) + 0(\tfrac{2}{3}) + 0(\tfrac{1}{4}) \;\; = 0$$
$$z_3 = 0(1) \;\;\; + 0(0) \;\; + 0(0) \;\; + 0(0) \;\;\;\; = 0$$
$$z_4 = 0(0) \;\;\; + 0(1) \;\; + 0(0) \;\; + 0(0) \;\;\;\; = 0$$
$$z_5 = 0(0) \;\;\; + 0(0) \;\; + 0(1) \;\; + 0(0) \;\;\;\; = 0$$
$$z_6 = 0(0) \;\;\; + 0(0) \;\; + 0(0) \;\; + 0(1) \;\;\;\; = 0.$$

We see that since the initial basic feasible solution consists entirely of slack variables, and since the c_j values for these variables are all zero, reducing the value of these slack variables when a nonbasic variable is introduced into solution causes no decrease in profit.

Since the objective function coefficient for x_1 is 10, the value of $c_1 - z_1$ is $10 - 0 = 10$, which indicates that the net result of bringing a unit of x_1 into the current solution will be an increase in profit of \$10. Hence, in the net evaluation row corresponding to x_1, we enter the value of 10.

In the same manner, we can calculate the corresponding z_j and $c_j - z_j$ values for the remaining variables. The result is the following complete initial Simplex tableau.

BASIS	c_j	x_1 10	x_2 9	s_1 0	s_2 0	s_3 0	s_4 0	
s_1	0	7/10	1	1	0	0	0	630
s_2	0	1/2	5/6	0	1	0	0	600
s_3	0	1	2/3	0	0	1	0	708
s_4	0	1/10	1/4	0	0	0	1	135
z_j		0	0	0	0	0	0	0
$c_j - z_j$		10	9	0	0	0	0	

PROFIT

In this tableau we also see a 0 in the z_j row in the last column. This zero represents the profit associated with the current basic solution. This value was obtained by multiplying the values of the basic variables, which are given in the last column, times their corresponding contribution to profit as given in the c_j column.

By looking at the net evaluation row, we see that every Standard bag Par produces will increase the value of the objective function by $10, and every Deluxe bag will increase the value of the objective function by $9. Given only this information it would make sense to produce as many Standard bags as possible. We know that every Standard bag produced uses $7/10$ of an hour of cutting and dyeing time; therefore, if we produce x_1 Standard bags, we will use $7/10 x_1$ hours of cutting and dyeing time. Since we only have 630 hours of cutting and dyeing time available, the maximum possible value of x_1, considering the cutting and dyeing constraint, can be calculated by solving the equation

$$7/10 x_1 = 630.$$

Thus, there is only enough time available in the cutting and dyeing department to manufacture a maximum of 900 Standard bags.

In a similar manner, every Standard bag produced uses ½ of an hour of the available 600 hours of sewing time; therefore, the maximum number of Standard bags we can produce and still satisfy the sewing constraint is given by

$$\tfrac{1}{2}x_1 = 600.$$

This indicates that x_1 could be at most 1200. But we know that it is impossible to produce 1200 Standard bags since we do not have enough cutting and dyeing time available. In fact, we saw that we only have enough capacity in the cutting and dyeing department to make 900 Standard bags. Considering these constraints simultaneously, the cutting and dyeing time is more restrictive. From the finishing constraint, we see that x_1 Standard bags would use $1x_1$ of the available 708 hours of finishing time. Solving the equation

$$1x_1 = 708$$

shows that in terms of the three constraints considered so far, we can produce at most 708 Standard bags.

In the inspection and packaging department, every Standard bag produced uses $\tfrac{1}{10}$ of an hour of inspection and packaging time. Since there is only 135 hours available we can solve

$$\tfrac{1}{10}x_1 = 135$$

to find that the largest number of Standard bags that can be processed by the inspection and packaging department is 1350. Thus, when we consider all the constraints together, we see that the most restrictive constraint in terms of the maximum number of Standard bags we can produce is the finishing constraint. That is, making 708 Standard bags will use all of the finishing capacity available. Hence, if x_1 is introduced into solution at its maximum value, we will produce 708 Standard bags ($x_1 = 708$) and there will be no slack time in the finishing department ($s_3 = 0$).

In making our decision to produce as many Standard bags as possible, we have changed the set of variables in our basic feasible solution. The previous nonbasic variable x_1 is now a basic variable with $x_1 = 708$, while the previous basic variable s_3 is now a nonbasic variable with $s_3 = 0$. This interchange of roles between two variables is the essence of the Simplex method. That is, the way the Simplex method moves from one basic feasible solution to another is by selecting a nonbasic variable to replace one of the current basic variables. This process of moving from one basic feasible solution to another is called an *iteration*.

Before presenting general rules for carrying out the steps of the Simplex method, let us consider the following constraint equation which might appear in the Tableau form of a linear program.

$$-\tfrac{2}{3}x_1 + 0x_3 \qquad + 1s_2 = 500.$$

Suppose that s_2 is a basic variable and x_1 and x_3 are non-basic variables. Since the coefficient of x_1 is negative ($-\tfrac{2}{3}$), every unit of x_1 introduced into solution would require the basic variable s_2 to increase by $\tfrac{2}{3}$ of a unit in order to maintain the constraint equation. Thus, no matter how large we make x_1, the basic variable s_2 will also become larger and hence will never be driven out of the basic solution (i. e., forced to zero). Similarly, since the coefficient of x_3 is zero, making x_3 basic would not affect the value of s_2. No matter how large we make x_3, the basic variable s_2 would remain unchanged and could never be driven out of solution. Thus, if the coefficient of a nonbasic variable is less than or equal to zero in some constraint, then that constraint can never limit the number of units of the non-basic variable that can be brought into solution. Hence, the basic variable associated with that constraint can never be driven out of solution. Therefore, in determining which variable should leave the current basis we only need to consider rows of our tableau in which the coefficient of the incoming nonbasic variable is *strictly positive*. With this additional consideration in mind, we now present the general Simplex rules for selecting a nonbasic variable to enter the basis and a current basic variable to leave the basis.

Criterion for Entering a New Variable into the Basis

Look at the net evaluation row and select as the variable to enter the basis that variable which will cause the largest per unit increase in the objective function. Let us say this variable corresponds to column j in the A portion of the tableau.

Criterion for Removing a Variable From the Current Basis

For each row i, compute the ratio b_i/a_{ij} for every a_{ij} greater than zero. This ratio tells us the maximum amount of variable x_j that can be brought into solution and still satisfy the constraint equation represented by that row. The minimum of these ratios tells us which constraint will be most restrictive if x_j is introduced into the solution. Hence we get the following rule for selecting the variable to remove from the current basis. *For all the ratios b_i/a_{ij} where $a_{ij} > 0$ select the basic variable corresponding to the minimum value of these ratios as the variable to leave the basis.*

Let us illustrate the above procedure by applying it to our Par, Inc. example problem. For illustration purposes, we have added an extra column showing the b_i/a_{ij} ratios for the initial Simplex tableau associated with our Par, Inc. problem.

BASIS	c_j	x_1 10	x_2 9	s_1 0	s_2 0	s_3 0	s_4 0		$\dfrac{b_i}{a_{i1}}$	
s_1	0	7/10	1	1	0	0	0	630	$\dfrac{630}{7/10} =$	900
s_2	0	1/2	5/6	0	1	0	0	600	$\dfrac{600}{1/2} =$	1200
s_3	0	①	2/3	0	0	1	0	708	$\dfrac{708}{1} =$	708
s_4	0	1/10	1/4	0	0	0	1	135	$\dfrac{135}{1/10} =$	1350
z_j		0	0	0	0	0	0	0		
$c_j - z_j$		10	9	0	0	0	0			

We see that $(c_1 - z_1) = 10$ is the largest positive value in the $(c_j - z_j)$ row. Hence, x_1 is selected to become the new basic variable. Checking the ratios b_i/a_{i1} for $a_{i1} > 0$, we see that $b_3/a_{31} = 708$ is the minimum of these ratios. Thus, the current basic variable associated with row 3 (s_3) is the variable selected to leave the basis. In our tableau we have circled a_{31} to indicate that the variable corresponding to the first column is to enter the basis and to indicate that the basic variable corresponding to the third row is to leave the basis. Adopting the usual linear programming terminology, we refer to this circled element as the *pivot element*.

We now see that to improve the current solution of $x_1 = 0$, $x_2 = 0$, $s_1 = 630$, $s_2 = 600$ $s_3 = 708$ and $s_4 = 135$, we should increase x_1 to 708. This would call for production of 708 Standard bags at a corresponding profit of $7080 = $10 x 708 units. In doing so, we will use all of the available finishing capacity, and thus s_3 will be reduced to zero. Hence, x_1 will become the new basic variable replacing s_3 in the old basis.

2.5   Calculating the Next Tableau

We saw in the previous section that our initial basic feasible solution could be improved by introducing x_1 into the basis to replace s_3. Before we can determine if this new basic feasible solution can be improved upon, it will be necessary to develop the corresponding Simplex tableau.

Recall that our initial Simplex tableau is simply a table containing the coefficients of the Tableau form for our linear programming problem. Because of the special properties of the Tableau form representation of the problem, our initial Simplex tableau contained a unit column corresponding to each of the basic variables. Thus, the value of the basic variable with a "1" in row i could be found by simply reading the ith element of the last column in the Simplex tableau, b_i.

What we would like to do now is to formulate a new tableau in such a fashion that all of the columns associated with the new basic variables are unit columns, and such that the value of the basic variable in row i is given by b_i. Thus, we would like to make the column in our new tableau corresponding to x_1 look just like the column corresponding to s_3 in our original tableau. Hence, our goal is to get the column in our A matrix corresponding to x_1 to appear as

$$0$$
$$0$$
$$1$$
$$0.$$

The way in which we transform the Simplex tableau so that it still represents an equivalent system of constraint equations with the above properties is to employ the elementary row operations discussed in Chapter 1. You will recall that there were two row operations that we could perform on a system of equations and still retain an equivalent system: (1) we could multiply any row by a nonzero number, or (2) we could multiply any row by a nonzero number and add it to another row. By performing these row operations, we will be able to change the column for the variable entering the basis to a unit column and, at the same time, change the last column of the tableau so that it contains the values of the new basic variables. We emphasize that performing these operations will in no way affect the solution to our problem, since the feasible solutions to the constraint equations are not changed by these elementary row operations.

Clearly, many of the numerical values in our new Simplex tableau are going to change as the result of performing these row operations. However, we know that after the row operations are performed, the new Simplex tableau will still represent an equivalent system of equations. Nonetheless, because the elements in the new Simplex tableau will usually change as the result of the required row operations, our present method of referring to elements in the Simplex tableau may lead to confusion. Let us see why this is so.

Up to now, we have made no distinction between the A matrix and b vector for Tableau form and the corresponding portions of the Simplex tableau. Indeed, we showed you that the initial Simplex tableau was formed by properly placing the a_{ij} and b_i elements as given in Tableau form into the Simplex tableau. From now on we will refer to the portion of the Simplex tableau that initially contained the a_{ij} values with the symbol $\bar{A}$, and the portion of the tableau that initially contained the b_i values with the symbol $\bar{b}$. In terms of the Simplex tableau, elements in $\bar{A}$ will be denoted by $\bar{a}_{ij}$ and elements in $\bar{b}$ will be denoted by $\bar{b}_i$. We recognize that using this notation we will have $\bar{A} = A$ and $\bar{b} = b$ in our initial Simplex tableau. However, in subsequent Simplex tableaus this relationship will usually not hold. This notation will avoid any possible confusion when we wish to distinguish between the original constraint coefficient values

(a_{ij}) and right-hand side values (b_i) of Tableau form, and the Simplex tableau elements $\bar{a}_{ij}$ and $\bar{b}_i$.

Now let us illustrate the procedure for calculating the next tableau by returning to the Par, Inc. problem. Recall that our goal is to get the column in the $\bar{A}$ portion of the tableau corresponding to x_1 to appear as

$$\begin{bmatrix} \bar{a}_{11} \\ \bar{a}_{21} \\ \bar{a}_{31} \\ \bar{a}_{41} \end{bmatrix} = \begin{bmatrix} 0 \\ 0 \\ 1 \\ 0 \end{bmatrix}$$

Since we already have $\bar{a}_{31} = 1$ in the initial Simplex tableau, no row operations need to be performed on the third row of our tableau.

In order to set $\bar{a}_{11} = 0$, we multiply our pivot row (the row corresponding to the finishing constraint) by $(-\frac{7}{10})$ to obtain the equivalent equation

$$-\tfrac{7}{10}(x_1 + \tfrac{2}{3}x_2 + 0s_1 + 0s_2 + 1s_3 + 0s_4) = -\tfrac{7}{10}(708)$$

or

$$-\tfrac{7}{10}x_1 - \tfrac{14}{30}x_2 - 0s_1 - 0s_2 - \tfrac{7}{10}s_3 + 0s_4 = -495.6. \qquad (2.11)$$

Now let us consider the cutting and dyeing constraint equation which is

$$\tfrac{7}{10}x_1 + 1x_2 + 1s_1 + 0s_2 + 0s_3 + 0s_4 = 630. \qquad (2.12)$$

Let us add equation (2.11) to the cutting and dyeing constraint equation (2.12). Dropping the terms with zero coefficients and performing this addition we get

$$(\tfrac{7}{10}x_1 + 1x_2 + 1s_1) + (-\tfrac{7}{10}x_1 - \tfrac{14}{30}x_2 - \tfrac{7}{10}s_3) = 630 - 495.6$$

or

$$0x_1 + \tfrac{16}{30}x_2 + 1s_1 - \tfrac{7}{10}s_3 = 134.4. \qquad (2.13)$$

Since this is just a simple row operation, we will have an equivalent system of equations if equation (2.12) is replaced by equation (2.13). Making this substitution in our original Simplex tableau, we see that we have obtained a zero in the first position in the x_1 column (i. e., $\bar{a}_{11} = 0$).

BASIS	c_j	x_1	x_2	s_1	s_2	s_3	s_4	
		10	9	0	0	0	0	
	0	0	16/30	1	0	$-7/10$	0	134.4
	1/2		5/6	0	1	0	0	600
	1		2/3	0	0	1	0	708
	1/10		1/4	0	0	0	1	135
z_j								
$c_j - z_j$								

We still need to set the elements in the second row and the fourth row of the x_1 column equal to zero. Can you find a way to do this? Recall that we accomplished this result for row 1 by multiplying the pivot row by a nonzero constant $(-\frac{7}{10})$ and then adding the result to the first row. Note that our constant in this case was just the negative of the coefficient in the first row and x_1 column. Thus, to get the element in the second constraint corresponding to the x_1 column equal to zero, we multiply the pivot row by $(-\frac{1}{2})$ and then add this result to the second constraint. This gives us the result

$$(\tfrac{1}{2}x_1 + \tfrac{5}{6}x_2 + 1s_2) + (-\tfrac{1}{2}x_1 - \tfrac{1}{3}x_2 - \tfrac{1}{2}s_3) = 600 - 354,$$

which is equivalent to

$$(0x_1 + \tfrac{1}{2}x_2 + 0s_1 + 1s_2 - \tfrac{1}{2}s_3 + 0s_4) = 246.$$

This becomes the new representation of the second constraint equation in our Simplex tableau.

To obtain a zero in the $\bar{a}_{41}$ position, we just multiply our pivot row by $(-\frac{1}{10})$ and then add the result to the last row. The resulting constraint equation is

$$0x_1 + \frac{22}{120}x_2 + 0s_1 + 0s_2 - \frac{1}{10}s_3 + 1s_4 = 64.2 .$$

Placing these last two equations into our new tableau gives us the following Simplex tableau.

BASIS	c_j	x_1 10	x_2 9	s_1 0	s_2 0	s_3 0	s_4 0	
s_1	0	0	16/30	1	0	-7/10	0	134.4
s_2	0	0	1/2	0	1	-1/2	0	246
x_1	10	1	2/3	0	0	1	0	708
s_4	0	0	22/120	0	0	-1/10	1	64.2
z_j								7080
$c_j - z_j$								

Since s_1, s_2, x_1, and s_4 are the basic variables in this tableau, x_2 and s_3 are set equal to 0, and we can read the solution for s_1, s_2, x_1 and s_4 directly from the tableau. That is,

$$s_1 = 134.4,$$
$$s_2 = 246,$$
$$x_1 = 708,$$
$$s_4 = 64.2 .$$

The profit corresponding to this solution is $7080. Note that this value for profit was obtained by multiplying the solution values for our basic variables in the $\bar{b}$ column times their corresponding objective function coefficients as given in the c_j column, $7080 = 0(134.4) + 0(246) + 10(708) + 0(64.2)$. We still have not calculated any entries in the z_j and $c_j - z_j$ rows. Before doing so, let us reflect for a moment on the present solution.

Interpreting the Results of an Iteration

Starting with one Simplex tableau, changing the basic variables, and finding a new Simplex tableau is referred to as an iteration of the Simplex method. In our example, the initial basic feasible solution was

$$\mathbf{x} = \begin{bmatrix} 0 \\ 0 \\ 630 \\ 600 \\ 708 \\ 135 \end{bmatrix}$$

with a corresponding profit of $0. One iteration of the Simplex method moved us to another basic feasible solution where the value of the objective function was $7080. This new basic feasible solution was

$$\mathbf{x} = \begin{bmatrix} 708 \\ 0 \\ 134.4 \\ 246 \\ 0 \\ 64.2 \end{bmatrix}$$

Graphically, this iteration moved us from one extreme point to another extreme point along the edge of our feasible region. In Figure 2–B we see that our initial basic feasible solution corresponded to extreme point 1. The first iteration moved us in the direction of the greatest per unit increase in profit, that is along the x_1 axis. We moved away from extreme point 1 in the x_1 direction until we could move no further without violating one of the constraints. The tableau we calculated after one iteration represents the basic feasible solution corresponding to extreme point 2.

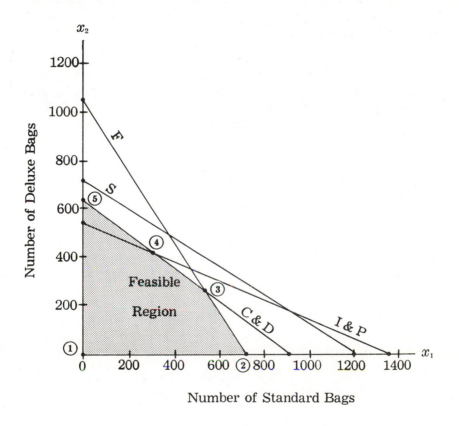

Figure 2–B. Feasible Solution Region for the Par, Inc. Problem

We know that the slack variables represent the unused capacity associated with each constraint. Noting the value of s_1 in our Simplex tableau, we see that the unused cutting and dyeing capacity is 134.4 hours. Does this seem reasonable? Well, since our solution indicated that we should make 708 Standard bags, and since each Standard bag requires $\frac{7}{10}$ of an hour of cutting and dyeing time, the total number of hours used in producing the 708 Standard bags is $(\frac{7}{10})$ (708) $= 495.6$. Since we started with 630 hours, we now have 134.4 hours of unused time available. Similarly, since every Standard bag produced requires $\frac{1}{2}$ of an hour of sewing time, the total amount of sewing time used in producing 708 Standard bags is 354 hours. We started with 600 hours of sewing time; therefore 246 hours remain. Every Standard bag requires 1 hour of finishing time. Thus since 708 hours of finishing time are available, we will use all of the finishing time by producing 708 Standard bags. This is the reason you see that the finishing constraint is binding at extreme point 2. Producing 708 Standard bags will use $70.8 =$ $(\frac{1}{10})$ 708 hours of inspection and packaging time leaving a slack of 64.2 hours in this department.

Moving Toward a Better Solution

We are ready to start things all over again. The next question we must ask ourselves is, can we find a new basic feasible solution (extreme point) that will increase the value of the objective function any further? To answer this, we need to calculate our z_j and $c_j - z_j$ rows for the current Simplex tableau.

Recall that the elements in the z_j row can be calculated by multiplying the elements in the c_j column of the Simplex tableau by the corresponding elements in the columns of the $\overline{A}$ matrix and summing. Doing this we get:

$$z_1 = 0(0) \quad + 0(0) \quad + 10(1) + 0(0) \quad = 10$$
$$z_2 = 0(\tfrac{16}{30}) \quad + 0(\tfrac{1}{2}) \quad + 10(\tfrac{2}{3}) + 0(\tfrac{22}{120}) = \tfrac{20}{3}$$
$$z_3 = 0(1) \quad + 0(0) \quad + 10(0) + 0(0) \quad = 0$$
$$z_4 = 0(0) \quad + 0(1) \quad + 10(0) + 0(0) \quad = 0$$
$$z_5 = 0(-\tfrac{7}{10}) + 0(-\tfrac{1}{2}) + 10(1) + 0(-\tfrac{1}{10}) = 10$$
$$z_6 = 0(0) \quad + 0(0) \quad + 10(0) + 0(1) \quad = 0.$$

Subtracting z_j from c_j to obtain the net evaluation row, we get the complete Simplex tableau.

BASIS	c_j	x_1 10	x_2 9	s_1 0	s_2 0	s_3 0	s_4 0	
s_1	0	0	16/30	1	0	−7/10	0	134.4
s_2	0	0	1/2	0	1	−1/2	0	246
x_1	10	1	2/3	0	0	1	0	708
s_4	0	0	22/120	0	0	−1/10	1	64.2
z_j		10	20/3	0	0	10	0	7080
$c_j - z_j$		0	7/3	0	0	−10	0	

Before considering the question of changing the basis and moving on to an even better basic feasible solution, let us see if we can interpret some of the numerical values appearing in the above Simplex tableau in terms of the original Par, Inc. problem.

We know that the elements in the x_2 column indicate how much each of the four basic variables will have to change in order to produce one unit of x_2 and still satisfy all the constraint relationships. Using the BASIS column to identify the basic variable corresponding to each element in the x_2 column, we can see that introducing one unit of x_2 will force us to decrease s_1 by $16/30$ of a unit, s_2 by ½ of a unit, x_1 by ⅔ of a unit, and s_4 by $22/120$ of a unit.

Why does producing one Deluxe bag require us to decrease the production of Standard bags by ⅔ of a unit? Notice that when we decided to produce 708 Standard bags we used all the finishing time available. Since every unit of x_2 we produce requires ⅔ of an hour of finishing time ($a_{32} = ⅔$), and every unit of x_1 requires 1 full hour, we see that in order to produce a unit of x_2 we will have to cut back ⅔ of a unit of x_1 in order to free up enough finishing time to produce a unit of x_2. Thus $\overline{a}_{32} = ⅔$ does indeed indicate correctly how many units of the basic variable x_1 must be given up if one unit of x_2 is introduced.

In our original tableau (LOOK BACK AND CHECK THIS) we saw that each Deluxe bag required *1 hour* of cutting and dye-

ing time.   Why then is $\bar{a}_{12} = {}^{16}\!/_{30}$?   Once again, each Deluxe bag
we produce will kick out $\frac{2}{3}$ of a Standard bag from the solution,
and hence free up $\frac{2}{3}$ of the cutting and dyeing time required
for one Standard bag.   Since each Standard bag requires $\frac{7}{10}$ of
an hour, we see that $(\frac{2}{3})$ $(\frac{7}{10})^| = {}^{14}\!/_{30}$ of an hour would be made
available because $\frac{2}{3}$ of a Standard bag leaves the solution.   Since
each Deluxe bag requires 1 hour of cutting and dyeing time, the
net effect of producing one Deluxe bag is to really only use up an
additional $(1-{}^{14}\!/_{30})$, or ${}^{16}\!/_{30}$ of an hour of cutting and dyeing time.
The remaining coefficients in the x_2 column can be interpreted
in the same manner.

To see why $(c_2 - z_2) = \frac{7}{3}$ we note that since the basic variables s_1, s_2 and s_4 are slack variables and have zero objective function coefficients, their reduction when one unit of x_2 is brought
into solution does not decrease total profit.   However, since the
profit associated with each unit of x_1 is \$10, the $\frac{2}{3}$ reduction will
cost us \$${}^{20}\!/_{3}$.   On the other hand, every unit of x_2 we bring into
solution increases profit by \$9, or \$${}^{27}\!/_{3}$.   Thus the net increase in
the value of the objective function resulting from a one unit in-
crease in x_2 will be given by \$${}^{27}\!/_{3} - {}^{20}\!/_{3} = \$\frac{7}{3}$.

Note that for all of the basic variables s_1, s_2, x_1, and s_4 the
value of $(c_j - z_j)$ is equal to zero.   Since each of these varia-
bles is associated with a unit column in our Simplex tableau,
we can interpret this as meaning that bringing one unit of a
basic variable into solution would force us to remove one unit
of the same basic variable.   The result obviously is no net change
in the value of the objective function.   From a physical point
of view the argument is even simpler.   Consider, for example,
x_1.   Since we are producing all we can of Standard bags, no
additional profit can be realized from their further production.
Thus, the net improvement in the objective function must be zero.

In summary we note that at each iteration of the Simplex
method

1.   the value of the current basic feasible solution can be
     found in the $\bar{b}$ column of the Simplex tableau;

2.   the value of $(c_j - z_j)$ for each of the basic variables is
     equal to zero;

3.   the coefficients in a particular column of the $\overline{A}$ portion
     of the Simplex tableau indicate how much the current

basic solution will change if one unit of the variable associated with that column is introduced.

Let us now analyze the net evaluation row to see if we can introduce a new variable into the basis and continue to improve the objective function. Using our rule for determining which variable should enter the basis next, we select x_2 since it has the highest positive coefficient in the net evaluation row.

In order to determine which variable will be removed from the basis when x_2 enters, we must compute for each row i the ratio $\dfrac{\bar{b}_i}{\bar{a}_{i2}}$ (remember though, we only compute this ratio if $\bar{a}_{i2}$ is greater than zero) and then select the variable corresponding to the minimum ratio as the variable to leave the basis. As before, we will show these ratios in an extra column of the Simplex tableau. Thus, our extended tableau becomes

BASIS	c_j	x_1 10	x_2 9	s_1 0	s_2 0	s_3 0	s_4 0		$\dfrac{\bar{b}_i}{\bar{a}_{i2}}$	
s_1	0	0	(16/30)	1	0	$-7/10$	0	134.4	$\dfrac{134.4}{16/30} =$	252
s_2	0	0	1/2	0	1	$-1/2$	0	246	$\dfrac{246}{1/2} =$	492
x_1	10	1	2/3	0	0	1	0	708	$\dfrac{708}{2/3} =$	1062
s_4	0	0	22/120	0	0	$-1/10$	1	64.2	$\dfrac{64.2}{22/120} =$	350.18
z_j		10	20/3	0	0	10	0	7080		
$c_j - z_j$		0	7/3	0	0	-10	0			

Since 252 is the minimum ratio, s_1 will be the variable that will leave the basis. Our pivot element will be $\bar{a}_{12} = {}^{16}\!/_{30}$, which is circled in the above tableau. The variable x_2 must now be made a basic variable. This means we must perform the operations necessary to convert the x_2 column into a unit column.

That is, we will have to transform the second column in our tableau to the form

$$\begin{bmatrix} 1 \\ 0 \\ 0 \\ 0 \end{bmatrix}$$

We can do this by performing the following steps:

Step 1.　Multiply every element in row 1 by $^{30}/_{16}$. Note that this will get us a 1 in position $\overline{a}_{12}$.

Step 2.　Multiply the new row 1 by $(-\frac{1}{2})$ and add the result to row 2. This will set $\overline{a}_{22} = 0$.

Step 3.　Multiply the new row 1 by $(-\frac{2}{3})$ and add the result to row 3. This will set $\overline{a}_{32} = 0$.

Step 4.　Multiply the new row 1 by $(-^{22}/_{120})$ and add the result to row 4. This will give us a zero in position $\overline{a}_{42}$.

The above elementary row operations again change the appearance of our Simplex tableau, but do not alter the solutions to the system of equations contained in the tableau. The only difference is that now we have x_2, s_2, x_1, and s_4 as the basic variables, and s_1 and s_3 as the nonbasic variables. The new tableau resulting from these row operations is presented below.

BASIS	c_j	x_1 10	x_2 9	s_1 0	s_2 0	s_3 0	s_4 0	
x_2	9	0	1	30/16	0	−210/160	0	252
s_2	0	0	0	−15/16	1	25/160	0	120
x_1	10	1	0	−20/16	0	300/160	0	540
s_4	0	0	0	−11/32	0	45/320	1	18
z_j								7668
$c_j - z_j$								

Note that the profit corresponding to this basic feasible solution is $252(9) + 120(0) + 540(10) + 18(0)$, or 7668, and that the basic variables are $x_2 = 252$, $s_2 = 120$, $x_1 = 540$, and $s_4 = 18$.

This basic feasible solution corresponds to extreme point 3 (see Figure 2–B). As you may recall from the graphical solution in Chapter 1, this is the optimal solution to the Par, Inc. problem. However, the Simplex method has not yet identified this solution as optimal. Thus we must continue to investigate whether or not it makes sense to bring any other variable into the basis and move to another basic feasible solution. As we saw before, this involves calculating the z_j and $c_j - z_j$ rows, and then selecting the variable to enter the basis that corresponds to the highest positive value in the net evaluation row.

After performing the z_j and $c_j - z_j$ calculations for our current solution, we obtain the following complete Simplex tableau.

BASIS	c_j	x_1 10	x_2 9	s_1 0	s_2 0	s_3 0	s_4 0	
x_2	9	0	1	30/16	0	$-210/160$	0	252
s_2	0	0	0	$-15/16$	1	25/160	0	120
x_1	10	1	0	$-20/16$	0	300/160	0	540
s_4	0	0	0	$-11/32$	0	45/320	1	18
z_j		10	9	70/16	0	111/16	0	7668
$c_j - z_j$		0	0	$-70/16$	0	$-111/16$	0	

Looking at the net evaluation row we see that every element is zero or negative. Since $(c_j - z_j)$ is less than or equal to zero for both of our nonbasic variables (s_1 and s_3), if we attempt to bring a nonbasic variable into the basis at this point, it will result in lowering the current value of the objective function. Hence, the above tableau represents the optimal solution to our linear programming problem.

Stopping Criterion

The optimal solution to a linear programming problem has been reached when there are no positive values in the net evaluation row of the Simplex tableau. If all entries in the net evaluation row are zero or negative, we stop the calculations and our optimal solution is given by the current Simplex tableau.

Interpreting the Optimal Solution

We see that in the final solution to our Par, Inc. problem the basic variables are x_2, s_2, x_1, and s_4. The complete optimal solution to the Par, Inc. problem is thus given by the following vector:

$$\mathbf{x} = \begin{bmatrix} 540 \\ 252 \\ 0 \\ 120 \\ 0 \\ 18 \end{bmatrix}$$

That is, our optimal solution is $x_1 = 540$, $x_2 = 252$, $s_1 = 0$, $s_2 = 120$, $s_3 = 0$, and $s_4 = 18$, with a corresponding value of the objective function of \$7668. Thus, if the management of Par, Inc. wants to maximize profit, they should produce 540 Standard bags and 252 Deluxe bags. In addition, management should note that there will be 120 hours of idle time in the sewing department and 18 hours of idle time in the inspection and packaging department. If it is possible to make alternate use of these additional resources management should plan to do so.

You can also see that with $s_1 = 0$ and $s_3 = 0$ there is no slack time available in the cutting and dyeing and the finishing departments. The constraints for these operations are both binding in our optimal solution (see Figure 2–B). If it is possible to obtain addititonal man-hours for these two departments, management should consider doing so.

2.6   Solution of a Sample Problem

In this section we will carry out the complete solution of a numerical example using the Simplex method. This is to provide you with an opportunity to check your understanding of the previous sections. You should attempt to solve the problem yourself on separate paper before studying the solution presented here.

The Problem

Solve the following linear program using the Simplex method.

$$\max \ 4x_1 + 6x_2 + 3x_3 + \ x_4$$
$$\text{s.t.}$$
$$\tfrac{3}{2}x_1 + 2x_2 + 4x_3 + 3x_4 \le 550$$
$$4x_1 + \ x_2 + 2x_3 + \ x_4 \le 700$$
$$2x_1 + 3x_2 + \ x_3 + 2x_4 \le 200$$
$$x_1, x_2, x_3, x_4 \ge 0$$

First we convert the problem to Standard form. Standard form for this problem is

$$\max \ 4x_1 + 6x_2 + 3x_3 + \ x_4 + 0s_1 + 0s_2 + 0s_3$$
$$\text{s.t.}$$
$$\tfrac{3}{2}x_1 + 2x_2 + 4x_3 + 3x_4 + 1s_1 \qquad\qquad = 550$$
$$4x_1 + \ x_2 + 2x_3 + \ x_4 \qquad + 1s_2 \qquad = 700$$
$$2x_1 + 3x_2 + \ x_3 + 2x_4 \qquad\qquad + 1s_3 = 200$$
$$x_1, x_2, x_3, x_4, s_1, s_2, s_3 \ge 0.$$

The next step is to write the problem in Tableau form. But, since all of the constraints are of the less-than-or-equal-to type and since the right-hand side values are all nonnegative, Standard form and Tableau form are the same. From this Tableau form we can set up the initial Simplex tableau.

BASIS	c_j	x_1 4	x_2 6	x_3 3	x_4 1	s_1 0	s_2 0	s_3 0		$\dfrac{\bar{b}_i}{\bar{a}_{i2}}$
s_1	0	3/2	2	4	3	1	0	0	550	$\dfrac{550}{2} = 225$
s_2	0	4	1	2	1	0	1	0	700	$\dfrac{700}{1} = 700$
s_3	0	2	③	1	2	0	0	1	200	$\dfrac{200}{3} = 66\tfrac{2}{3}$
z_j		0	0	0	0	0	0	0	0	
$c_j - z_j$		4	6	3	1	0	0	0		

Note that all entries in the z_j row are zero because all of the basic variables are slack variables, and the entries in the z_j row are just the result of multiplying each element in the c_j column by the corresponding element in column j and summing these products. For example $z_3 = 0(4) + 0(2) + 0(1) = 0$.

Since $(c_2 - z_2) = 6$ has the largest positive value in the net evaluation row, we select variable x_2 to enter the basis. Calculating the ratios $\bar{b}_i/\bar{a}_{i2}$, we see that the minimum ratio is $\bar{b}_3/\bar{a}_{32} = 66\tfrac{2}{3}$. Hence, the third row becomes our pivot row and $\bar{a}_{32} = 3$ is the pivot element, which is circled in our Simplex tableau.

In order to obtain a 1 in the $\bar{a}_{32}$ position we multiply the pivot row by ($\tfrac{1}{3}$). The following partial Simplex tableau results.

BASIS	c_j	x_1 4	x_2 6	x_3 3	x_4 1	s_1 0	s_2 0	s_3 0	
		3/2	2	4	3	1	0	0	550
		4	1	2	1	0	1	0	700
		2/3	1	1/3	2/3	0	0	1/3	66⅔
z_j									
$c_j - z_j$									

We want to make the x_2 column a unit column so we must perform the necessary row operations to obtain zeros for $\bar{a}_{12}$ and $\bar{a}_{22}$. We set $\bar{a}_{12} = 0$ by multiplying the new pivot row by (-2) and adding the result to row 1. The result of this operation is shown below.

		x_1	x_2	x_3	x_4	s_1	s_2	s_3	
BASIS	c_j	4	6	3	1	0	0	0	
		1/6	0	10/3	5/3	1	0	-2/3	416⅔
		4	1	2	1	0	1	0	700
		2/3	1	1/3	2/3	0	0	1/3	66⅔
	z_j								
	$c_j - z_j$								

Finally, we set $\bar{a}_{22}$ equal to zero by multiplying the new pivot row by (-1) and adding it to row 2. After performing this operation we can present the complete Simplex tableau for our new basic feasible solution. The basic variables in this solution are s_1, s_2 and x_2.

		x_1	x_2	x_3	x_4	s_1	s_2	s_3		$\dfrac{\bar{b}_i}{\bar{a}_{i3}}$
BASIS	c_j	4	6	3	1	0	0	0		
s_1	0	1/6	0	10/3	5/3	1	0	-2/3	416⅔	125
s_2	0	10/3	0	5/3	1/3	0	1	-1/3	633⅓	380
x_2	6	2/3	1	1/3	2/3	0	0	1/3	66⅔	200
	z_j	12/3	6	6/3	12/3	0	0	6/3	400	
	$c_j - z_j$	0	0	3/3	-9/3	0	0	-6/3		

We see that our current solution is not optimal since $(c_3 - z_3) = ⅔$ is positive. This means that every unit of x_3 introduced into solution will cause our objective function to increase by one. Since $(c_3 - z_3)$ is the largest positive entry in the net evaluation row, we select x_3 to be introduced into solution at our next iteration. The smallest ratio of b_i / a_{i3} is found to be $b_1 / a_{13} = 416⅔ / (10/3)$

$= 125$; therefore, row 1 becomes our new pivot row. The row operations necessary to transform the x_3 column into a unit column are summarized below.

Step 1. Multiply row 1 by ($3/10$). This is the new pivot row and we now have $\overline{a}_{13} = 1$ as desired.

Step 2. Multiply the new pivot row by ($-5/3$) and add the result to row 2. This gives us a zero in the $\overline{a}_{23}$ position.

Step 3. Multiply the new pivot row by ($-1/3$) and add the result to row 3. This gives us a zero in the $\overline{a}_{33}$ position.

The Simplex tableau for our new basic feasible solution may now be completed. This completed tableau is presented as follows.

BASIS	c_j	x_1	x_2	x_3	x_4	s_1	s_2	s_3	
		4	6	3	1	0	0	0	
x_3	3	3/60	0	1	5/10	3/10	0	$-2/10$	125
s_2	0	39/12	0	0	$-15/30$	$-5/10$	1	0	425
x_2	6	39/60	1	0	15/30	$-1/10$	0	12/30	25
z_j		81/20	6	3	9/2	3/10	0	54/30	525
$c_j - z_j$		$-1/20$	0	0	$-7/2$	$-3/10$	0	$-54/30$	

All of the $(c_j - z_j)$ elements are less than or equal to zero; hence, there is no variable which we can introduce into the solution and obtain an increase in the objective function. Therefore, the current solution is optimal. The basic variables are x_3, s_2 and x_2, and the complete optimal solution vector is given by

$$\mathbf{x} = \begin{bmatrix} 0 \\ 25 \\ 125 \\ 0 \\ 0 \\ 425 \\ 0 \end{bmatrix}$$

The value of the objective function for this solution is 525.

2.7   Summary

In Chapter 1 we saw how small linear programs could be solved using a graphical approach. In this chapter the Simplex method was developed as a procedure for solving larger linear programs. Actually the Simplex method is also an easy way to solve small linear programs by hand calculations as well. However, as problems get larger even the Simplex method becomes too cumbersome for efficient hand computation. As a result, we must utilize the computer if we want a solution to larger linear programs in any reasonable length of time. If you have access to a computer, you may want to consider using the linear programming computer code provided in Appendix A. The necessary instructions for using the code are also included along with an example problem.

As a review of the material in this chapter we present here a detailed step-by-step procedure for solving linear programs using the Simplex method.

Step	Justification
1. Formulate a linear programming model of the real-world problem.	This is to obtain a mathematical representation of the problem.
2. Set up the Standard form representation of the linear program.	This is to make every constraint an equality and is the first step in preparing the problem for solution using the Simplex method.
3. Set up the Tableau form representation of the linear program.	This is necessary in order to obtain an initial basic feasible solution. All linear programs must be put in this form before the initial Simplex tableau can be set up.
4. Set up the Simplex tableau.	This will be used to keep track of the calculations made as we carry out the Simplex method. The solution corresponding to the initial Simplex tableau is always the origin.

Step	Justification
5. Choose the variable with the largest $(c_j - z_j)$ to introduce into solution.	The value of $(c_j - z_j)$ tells us the amount by which the value of the objective function will increase for every unit of x_j introduced into solution.
6. Choose as the pivot row that row with the smallest ratio of $\bar{b}_i / \bar{a}_{ij}$, $\bar{a}_{ij}$ greater than zero.	This determines which variable will leave the basis when x_j enters. This also tells us how many units of x_j can be introduced into solution before the basic variable in the ith row equals zero.
7. Perform the necessary row operations to convert column j to a unit column.	Once these row operations have been performed we can read the values of our basic variables from the $\bar{b}$ column of our tableau.
7a. Multiply the pivot row by the constant necessary to make the pivot element a 1.	
7b. Obtain zeros in all the other rows by multiplying the new pivot row by an appropriate constant and adding it to the appropriate row.	
8. Test for optimality. If $(c_j - z_j) \leq 0$ for all columns we have the optimal solution. If not, we return to step 5.	If $(c_j - z_j) \leq 0$ for all variables this means there is no variable that we can introduce which will cause the objective function to increase.

Some additional notation was introduced in this chapter. We use $\bar{A}$ and $\bar{a}_{ij}$ to denote the positions in the Simplex tableau corresponding to the A matrix in the Tableau form representation of our linear program. Similarly, we use $\bar{b}$ and $\bar{b}_i$ to denote the positions in the Simplex tableau corresponding to the b vector in our Tableau form representation.

2.8   Glossary

1. *Basic solution*—For a general linear program with n-variables and m-constraints a basic solution may be found by setting (n-m) of the variables equal to zero and solving the constraint equations for the values of the other m-variables.   If a unique solution exists, it is a basic solution.

2. *Basic feasible solution*—This is simply a basic solution which is also in the feasible region (i. e., it satisfies the nonnegativity requirement).   A basic feasible solution corresponds to an extreme point.

3. *Unit vector or unit column*—This is a vector, or column of a matrix, which has a zero in every position except one. In the nonzero position there is a "1".

4. *Tableau form*—This is the form a linear program must be written in prior to setting up the initial Simplex tableau. When a linear program is written in this form its matrix representation contains m unit columns corresponding to basic variables and the values of these basic variables are given by the b vector.   A further requirement is that the entries in the b vector be greater than or equal to zero. This requirement provides us with a basic feasible solution.

5. *Simplex method*—This is an algebraic procedure for solving linear programs.   Basically it moves from one basic feasible solution (extreme point) to another making sure that the objective function increases at each iteration until the optimal solution is reached.

6. *Simplex tableau*—This is a table used to keep track of the calculations made when the Simplex solution method is employed.

7. *Current solution*—When carrying out the Simplex method the current solution refers to the basic feasible solution (extreme point) that we are now at.

8. *Pivot row*—This is the row in the Simplex tableau corresponding to the basic variable which will leave the solution

as the algorithm iterates from one basic feasible solution to another.

9. *Pivot column*—This is the column corresponding to the non-basic variable which is about to be introduced into the basic feasible solution.

10. *Pivot element*—This is the element of the Simplex tableau that is in both the pivot row and pivot column.

11. *Net evaluation row*—This is the $(c_j - z_j)$ row. The jth element in this row indicates the amount the value of the objective function will increase if one unit of x_j is introduced into solution.

12. *Basis*—This is the set of variables which are not restricted to equal zero in the current basic solution. The variables which make up the basis are termed basic variables, and the remaining variables are called non-basic variables.

13. *Iteration*—An iteration of the Simplex method consists of the sequence of steps performed in moving from one basic feasible solution to another.

2.9 Problems

1. If the following partial initial Simplex tableau is given:

		x_1	x_2	x_3	s_1	s_2	s_3	
BASIS	c_j	5	20	25	0	0	0	
		2	1	0	1	0	0	40
		0	2	1	0	1	0	30
		3	0	-1/2	0	0	1	15
	z_j							
	$c_j - z_j$							

a. Complete the initial tableau.

b. Write the problem in its Tableau form.

c. What is the initial basis? Does this correspond to the origin? Explain.

d. What is the value of the objective function at this initial solution?

e. For the next iteration, what variable should enter the basis and what variable should leave the basis?

f. How many units of the entering variable will be in the next solution? Before making this first iteration, what should be the value of the objective function after the first iteration?

g. Find the optimal solution using the Simplex method.

2. Solve the following linear program using the graphical approach. Next set the linear program up in Tableau form and use the Simplex method to solve. Show on your graph the sequence of extreme points generated by the Simplex method.

$$\max\ x_1 + x_2$$

s.t.

$$\tfrac{1}{2}x_1 + x_2 \le 3$$
$$2x_1 + x_2 \le 5$$
$$x_1, x_2 \ge 0$$

3. Explain in your own words why Tableau form and Standard form are the same for problems with less-than-or-equal-to constraints and nonnegative b_i's.

4. Solve the Ryland Farms problem from Chapter 1 (No. 20) using the Simplex method. Compare each iteration to the graphical solution procedure.

5. Solve the Wilkinson Motors, Inc. problem from Chapter 1 (No. 19) using the Simplex method. Compare each iteration to the graphical solution procedure.

6. Solve the following linear program. What are the values of your basic variables at each iteration?

$$\max\ 3x_1 + 5x_2 + x_3 + 2x_4$$

s.t.

$$\tfrac{1}{2}x_1 + \tfrac{7}{10}x_2 + \tfrac{1}{10}x_3 + \tfrac{4}{10}x_4 \le 800$$
$$x_1 + x_2 + \tfrac{8}{10}x_3 + \tfrac{1}{2}x_4 \le 650$$
$$\tfrac{6}{10}x_1 + \tfrac{5}{10}x_2 + \tfrac{5}{10}x_3 + \tfrac{6}{10}x_4 \le 480$$
$$x_1, x_2, x_3, x_4 \ge 0$$

7. Solve the following linear program.

    max $5x_1 + 2x_2 + 8x_3$
    s.t.

    $$x_1 - 2x_2 + \tfrac{1}{2}x_3 \leq 420$$
    $$2x_1 + 3x_2 - x_3 \leq 610$$
    $$6x_1 - x_2 + 3x_3 \leq 125$$
    $$x_1,\ x_2,\ x_3 \geq 0$$

8. Solve the following linear program using both the graphical and the Simplex method. Show graphically how the Simplex method moves from one extreme point to another.

    max $x_1 + x_2$
    s.t.

    $$x_1 + 2x_2 \leq 6$$
    $$6x_1 + 4x_2 \leq 24$$
    $$x_1,\ x_2 \geq 0$$

    Find the coordinates of all the extreme points of this problem.

9. How many basic solutions are there to a linear program which has 8 variables and 5 constraints when written in Standard form?

10. Explain in your own words why it is that when we are trying to determine which basic variable to eliminate at a particular iteration we only consider $\bar{a}_{ij}$'s which are strictly greater than zero.

11. Find all of the basic solutions for the linear program in problem 8. (Hint: There are six.) Which of these are extreme points?

12. Suppose that instead of picking the variable with the largest positive value in the net evaluation row to introduce at each iteration, we introduce any variable with a positive value of $(c_j - z_j)$ without regard to whether or not it is the largest value. Do you think we would still reach the optimal solution? Why or why not?

13. Suppose that we did not remove the basic variable with the smallest ratio of $\bar{b}_i/a_{ij}$ at a particular iteration. What effect would this have on the Simplex tableau for our next solution?

14. Suppose a company manufactures three products from two raw materials where

	Product 1	Product 2	Product 3
Raw Material A	3.5 lbs.	3 lbs.	1.5 lbs.
Raw Material B	2.5 lbs.	2 lbs.	1 lb.

If the company has available 50 pounds of material A and 100 pounds of material B and if the profits for the three products are $10, $10, and $7.50, how much of each product should be produced in order to maximize profits?

15. Liva's Lumber, Inc. manufactures three types of plywood. The data below summarizes the production hours per unit in each of three production operations and other data for the problem.

	Operations			Profit
Plywood	I	II	III	per Unit
Grade A	2	2	4	40
Grade B	5	5	2	30
Grade X	10	3	2	20
Maximum Hours Available	900	400	600	

How many units of each grade of lumber should be produced?

16. Ye Olde Cording Winery in Peoria, Illinois makes three kinds of authentic German wine: Heidelberg Sweet, Heidelberg Regular, and Deutschland Extra Dry. The raw materials, labor, and profit for a gallon of each of these wines is summarized below:

Wine	Grapes Grade A (Bushels)	Grapes Grade B (Bushels)	Sugar (lbs.)	Man-hours	Profit per Gallon
Heidelberg Sweet	1	1	2	2	10
Heidelberg Regular	2	0	1	3	12
Deutschland Extra Dry	0	2	0	1	20

If the Winery has 150 bushels of grade A grapes, 150 bushels of grade B grapes, 80 pounds of sugar, and 225 man-hours available during the next week, what product mix of wines will maximize the company's profit?

a. Solve by the Simplex method.

b. Interpret all slack variables.

c. An increase in which resources could improve the company's profit?

17. The Our-Bags-Don't-Break (OBDB) Plastic Bag Company manufactures three plastic refuse bags for home use: a 20-gallon garbage bag, a 30-gallon garbage bag, and a 33-gallon leaf and grass bag. Using purchased plastic material, three operations are required to produce each end-product: cutting, sealing, and packaging. The production time required to process each type of bag in every operation, as well as the maximum daily production time available for each operation, is shown below. Note that the production time figures in this table are for a box of each type of bag.

Production Time (Seconds)

Type of Bag	Cutting	Sealing	Packaging
20-gallon	2	2	3
30-gallon	3	2	4
33-gallon	3	3	5
Time Available	2 hrs.	3 hrs.	4 hrs.

If OBDB makes a profit of $0.10 for each box of 20-gallon bags produced, $0.15 for each box of 30-gallon bags, and $0.20 for each box of 33-gallon bags, what is the optimal product mix?

18. Kirkman Brothers Ice Cream Parlors sell three different flavors of Dairy Sweet ice milk: chocolate, vanilla, and banana. Due to extremely hot weather and a high demand for its products, Kirkman has run short of its supply of ingredients: milk, sugar and cream. Hence, Kirkman will not be able to fill all of the orders received from its retail outlets (the Ice Cream Parlors). Due to these circumstances, Kirkman has decided to make the best amounts of the three flavors given the constraints on supply of the basic ingredients. The company will then ration the ice milk to the retail outlets.

Kirkman has collected the following data on profitability of the various flavors, availability of supplies and amounts required by each flavor.

	Profit per Gallon	Usage per Gallon		
		Milk	Sugar	Cream
Chocolate	$1.00	.45 gal	.50 lbs	.10 gal
Vanilla	$0.90	.50 gal	.40 lbs	.15 gal
Banana	$0.95	.40 gal	.40 lbs	.20 gal
Max Available		200 gal	150 lbs	60 gal

Determine the optimal product mix for Kirkman Brothers. What additional resources could be used?

19. Uforia Corporation sells two different brands of women's perfume: Incentive and Temptation No. 1. Uforia sells exclusively through department stores and employs a three-man sales staff to call on its customers.

The amount of sales time necessary for each of the salesmen to sell one case of each product varies with the experience and ability of the salesmen. Data on the average times for each of Uforia's salesmen is presented below.

	Average Sales Time per Case	
	Incentive	Temptation No. 1
John	10 mins.	15 mins.
Alex	15 mins.	10 mins.
Red	12 mins.	6 mins.

Each of the salesmen spends approximately 80 hours per month in the actual selling of these two products. Each case of Incentive and Temptation No. 1 sells at a profit of $30 and $25, respectively.

Uforia has a twofold problem. First it would like to know how many cases of each perfume to produce over the next month in order to maximize profit. Secondly, the Corporation would like to know how much of each salesman's time should be allocated to each of the products.

Use linear programming to solve the above problems for Uforia. (Hint: Let x_1 = number of cases of Incentive sold by John x_2 = number of cases of Temptation No. 1 sold by John, x_3 = number of cases of Incentive sold by Alex, etc.)

20. Solve problem 25 in Chapter 1 using the Simplex method.

21. Suppose the management at Par, Inc. learned the accounting department made a mistake and that the profit on the Deluxe bag was really $18 per bag. Shown below is a partial Simplex tableau corresponding to the optimal basic feasible solution with $c_1 = 10$ and $c_2 = 9$. The only difference is that we have changed c_2 to 18.

   a. Calculate the remainder of the Simplex tableau and show that the Simplex method indicates the current solution is not optimal.

   b. Find the optimal solution with $c_2 = 18$. What new variable enters the basis? What variable leaves?

   c. Refer to the original graphical solution in Figure 1–L. What extreme point is now optimal? What constraints are now binding?

		x_1	x_2	s_1	s_2	s_3	s_4	
BASIS	c_j	10	18	0	0	0	0	
x_2	18	0	1	30/16	0	−21/16	0	252
s_2	0	0	0	−15/16	1	25/160	0	120
x_1	10	1	0	−20/16	0	300/160	0	540
s_4	0	0	0	−11/32	0	45/320	1	18
	z_j							
	$c_j − z_j$							

22. Catalina Yachts, Inc. is a builder of cruising sail boats. They manufacture three models of sail boats: the C–32, the C–40 and the C–48. The company, because of its excellent reputation, is in the enviable position of being able to sell all the boats it manufactures. Catalina is currently in the process of taking orders for the coming year. The

company has asked you to determine for them how many orders for each model it should accept in order to maximize profits.

The manufacture of each model requires differing amounts of time spent on each of three operations: molding, carpentry and finishing. The number of days required to perform each of these activities on the three models is given below.

Model	Man/Days for Molding	Man/Days for Carpentry	Man/Days for Finishing
C–32	3	5	4
C–40	5	12	5
C–48	10	18	8

Based on past experience management expects the profit per boat to be: $5,000 on the C–32, $10,000 on the C–40, $20,000 on the C–48.

Catalina currently has 40 people employed in manufacturing these sail boats; 10 in molding, 20 in carpentry, 10 in finishing, and on the average each employee works 240 days per year. The only other constraint is a management imposed restriction on the number of C–48 models that can be sold. Because Catalina does not want the C–48 to become commonplace, it will not take orders for over 20 of this model.

23. The World-Wide Grocery Store Company, in preparation for the upcoming holiday season, has just purchased the following quantities of nuts.

Type of Nut	Amount (Pounds)
Almonds	6000
Brazil	7500
Filberts	7500
Pecans	4000
Walnuts	7500

They would like to package these nuts in one-pound bags, and are presently considering producing a regular mix (consisting of 10% Pecans, 15% Almonds, 25% Filberts, 25% Brazil, 25% Walnuts), a deluxe mix (20% of each type of nut), and individual bags of each type of nut. The profit figures for each bag they produce are as follows:

Type of Bag	Profit/Bag
Regular Mix	$0.20
Deluxe Mix	$0.25
Almonds	$0.05
Brazil	$0.10
Filberts	$0.10
Pecans	$0.05
Walnuts	$0.15

Your job is to formulate a linear program that World-Wide could use in order to determine how many bags of each type they should produce to maximize profits. Solve the linear program you formulate in order to obtain the optimal solution. Do so on the computer if one is available.

24. The employee credit union for Ivory Tower University is planning its usage of funds for the coming year. The credit union makes four different kinds of loans to members, each of which has a different rate of return. Also, in order to stabilize income, the company is permitted to invest up to 30% of its money in "risk-free" securities. The various revenue producing investments together with their annual return are as follows.

Type of Investment	Annual Rate of Return (%)
Automobile loans (secured)	8
Signature loans	12
Furniture loans (secured)	10
Other secured loans	11
Securities (risk free)	9

Credit union policy and state law impose the following restrictions on the composition of the above company assets.

Securities may not exceed 30% of the total amount of funds available for investment and signature loans may not exceed 10% of total loans. Furniture loans plus "other secured loans" may not exceed 50% of total secured loans (the total of all three types). Signature loans plus "other secured loans" may not exceed the amount invested in securities.

The credit union would like to determine how much of its funds should be allocated to each of the investments in order to maximize profits. The firm expects to have $3 million to invest in the coming year.

Formulate this as a linear program and solve for the optimal solution using either the computer code in the Appendix or your own code.

3

The Minimization Problem and Other Topics

All of the linear programming problems you have been introduced to so far have involved *maximizing* a linear function subject to a set of linear constraints. In practice, however, we often find problems where the objective becomes one of *minimization*. For example, a manufacturer wants to find a minimum cost production schedule, a trucking company wants to find minimum total distance shipping routes, etc. In addition, the only types of constraints you have encountered have been less-than-or-equal-to constraints. Frequently, many of the constraints associated with a linear programming problem take the form of greater-than-or-equal-to constraints and/or equality constraints. One of the major purposes of this chapter is to introduce you to these other forms of the linear programming problem. In addition, we shall also consider some special cases that will illustrate the various solution possibilities that can arise when we solve any linear programming problem. Let us begin by considering what changes we might need to make in our Simplex procedure in order to handle the greater-than-or-equal-to and/or the equality types of constraints.

3.1 Surplus Variables

Let us consider again the problem faced by Par, Inc. That is, what is the maximum profit mix of Standard and Deluxe bags? We previously saw that the mathematical statement of the problem involved all $\leq$ constraints, and was written as

$$\max \ z = 10x_1 + 9x_2$$

s.t.

$$\tfrac{7}{10}x_1 + 1x_2 \leq 630$$
$$\tfrac{1}{2}x_1 + \tfrac{5}{6}x_2 \leq 600$$
$$1x_1 + \tfrac{2}{3}x_2 \leq 708$$
$$\tfrac{1}{10}x_1 + \tfrac{1}{4}x_2 \leq 135$$
$$x_1, \ x_2 \geq 0 \ .$$

We have seen that from the optimal solution to this problem the management of Par, Inc. will be able to determine how many Standard and Deluxe bags should be produced in order to maximize total profit. We now note that since maximizing profits is Par's only criterion, it is entirely possible that the optimal solution to such a problem would indicate that the company should not produce any of one particular type of bag. Realizing this, the management of Par, Inc. might want to ensure that at least a certain number of each the Standard and Deluxe bags be produced. That is, in this initial planning period, management would probably want to assess the public reaction to both types of bags. Let us suppose, for example, that management wants to ensure that the optimal solution consists of at least one-hundred bags of each model.

How do these addititonal restrictions change the mathematical form of our problem? One way we can handle these new restrictions is by adding a constraint that ensures that x_1 will be greater than or equal to one-hundred bags, and adding another constraint that ensures that x_2 will be greater than or equal to one-hundred bags.[1] That is, we can add the constraints,

$$x_1 \geq 100$$
$$x_2 \geq 100.$$

------◆------

[1] In problem 20 at the end of this chapter an alternative procedure for handling lower bound constraints such as these is developed.

The only real difference in the mathematical form of our problem is that the last two constraints become $\geq$ constraints. With these two additions, our modified problem can now be written as

$$\max z = 10x_1 + 9x_2$$

s.t.

$$\tfrac{7}{10}x_1 + 1x_2 \leq 630$$

$$\tfrac{1}{2}x_1 + \tfrac{5}{6}x_2 \leq 600$$

$$1x_1 + \tfrac{2}{3}x_2 \leq 708$$

$$\tfrac{1}{10}x_1 + \tfrac{1}{4}x_2 \leq 135$$

$$1x_1 \qquad\qquad \geq 100$$

$$1x_2 \geq 100$$

$$x_1, x_2 \geq 0$$

For this particular problem we see that the additional constraints also guarantee that our variables are nonnegative; thus, we really could have eliminated writing $x_1, x_2 \geq 0$. However, in many problems where the $\geq$ constraints are not simply lower bounds on the decision variables, this will not be the case; therefore, we will always include the nonnegativity relationships, realizing that in some situations they would not be required.

Let us now see the effect these additional constraints have on the graphical solution to our Par, Inc. problem. In Figure 3–A we show the feasible region for the Par, Inc. problem after the addition of the two greater-than-or-equal-to constraints. The feasible region is the shaded portion of the figure.

We see that the addition of these two new constraints has the effect of making infeasible all solutions which have $x_1 < 100$ and $x_2 < 100$. Note, however, that although these new constraints have altered the feasible region, the optimal solution will still be at extreme point ③ where $x_1 = 540$ and $x_2 = 252$.

In terms of the graphical solution method, no change in the solution procedure is necessary. That is, simply find the highest value of z for which the feasible region and the objective function have a point in common. However, if we are to use the Simplex method for solving this problem, we must learn how to handle the greater-than-or-equal-to constraints in setting up

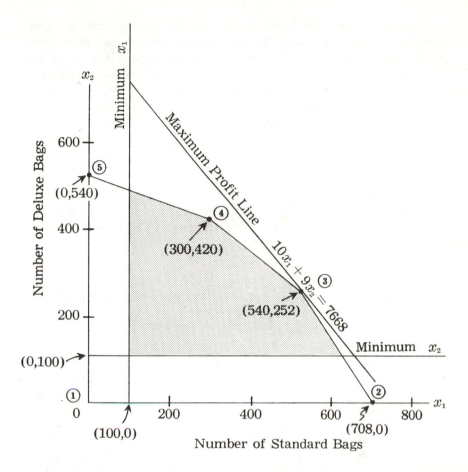

Figure 3–A. Feasible Region for the Modified
Par, Inc. Problem

the initial Simplex tableau. Hence, we must first learn to use
a procedure that will convert our ≥ constraints into the equality
constraints required by the Standard form representation of a
linear programming problem. Then, we must convert our Standard form representation into Tableau form.

Let us recall the method we used for converting a less-than-
or-equal-to constraint into an equality constraint. For example,
consider the cutting and dyeing constraint

$$\tfrac{7}{10}x_1 + 1x_2 \leq 630.$$

We reasoned that since the left-hand side of this relation was always less than or equal to the right-hand side, the only way an equality could ever exist would be if we added something to the left-hand side to make up for the difference between 630 and $(\frac{7}{10}x_1 + 1x_2)$. Since the left-hand side of the constraint varies depending on the values of x_1 and x_2, we concluded that the amount we added must also be a variable. Appropriately, we named this added variable a slack variable. Our equation then became

$$\frac{7}{10}x_1 + 1x_2 + s_1 = 630.$$

What change would we have to make if our original equation was of the greater-than-or-equal-to form. That is, what if

$$\frac{7}{10}x_1 + 1x_2 \geq 630.$$

If this were the case, the only way the left-hand and right-hand sides could be made equal would be if we subtracted some positive amount from the left-hand side to make up for the difference between $(\frac{7}{10}x_1 + 1x_2)$ and 630. (Or, equivalently, we could have added something to the right-hand side.) Since the left-hand side varies depending on the values of x_1 and x_2, the amount we subtract must also be a variable. Since there was "too much" on the left-hand side, we refer to the variable we will subtract as a surplus variable. Because surplus variables essentially serve the same purpose as slack variables, i. e., changing inequality constraints to equality constraints, we will also use the symbol s with an appropriate subscript to identify these variables. Subtracting the surplus variable s_1 our greater-than-or-equal-to constraint becomes

$$\frac{7}{10}x_1 + 1x_2 - s_1 = 630.$$

In our modified Par, Inc. example where management imposed additional restrictions, the greater-than-or-equal-to constraints were

$$x_1 \geq 100,$$

and

$$x_2 \geq 100.$$

In setting up our original Par, Inc. problem in Standard form, we defined slack variables s_1, s_2, s_3, and s_4 to make the first four constraints equalities. We will now define surplus variables s_5 and s_6 in order to write our greater-than-or-equal-to constraints as the equalities

$$x_1 - s_5 = 100,$$
and
$$x_2 - s_6 = 100.$$

These surplus variables (s_5 and s_6) will be interpreted as the surplus or excess production over the minimum production requirement of 100 units for both bags. For example, the optimal solution to this problem calls for the production of 540 Standard and 252 Deluxe bags. Thus there will be a surplus of 440 Standard and 152 Deluxe bags with respect to these constraints in the optimal solution.

Just as was the case for our slack variables, we recognize that these surplus variables make no contribution to profit; therefore, they are assigned a coefficient of zero in the objective function. We can now write our current problem in Standard form.

max $10x_1 + 9x_2 + 0s_1 + 0s_2 + 0s_3 + 0s_4 + 0s_5 + 0s_6$
s.t.

$$
\begin{array}{lll}
{}^{7}\!/_{10}x_1 + 1x_2 + 1s_1 & = 630 & (3.1) \\
\tfrac{1}{2} x_1 + {}^{5}\!/_{6}x_2 \quad +1s_2 & = 600 & (3.2) \\
1x_1 + {}^{2}\!/_{3}x_2 \quad +1s_3 & = 708 & (3.3) \\
{}^{1}\!/_{10}x_1 + \tfrac{1}{4}x_2 \quad +1s_4 & = 135 & (3.4) \\
1x_1 \quad - 1s_5 & = 100 & (3.5) \\
+ 1x_2 \quad - 1s_6 & = 100 & (3.6)
\end{array}
$$

$$x_1, x_2, s_1, s_2, s_3, s_4, s_5, s_6 \geq 0$$

3.2 Artificial Variables

Let us now consider again the way we generated an initial basic feasible solution to get the Simplex method started. We set $x_1 = 0$, $x_2 = 0$, and selected the slack variables as our initial basic variables. Extending this notion to our current problem would suggest setting $x_1 = 0$, $x_2 = 0$, and selecting as initial basic variables the slack and surplus variables. However, looking at Figure 3–A, we see that this solution, corresponding to the origin, is no longer feasible. The inclusion of the two greater-than-or-equal-to constraints, $x_1 \geq 100$ and $x_2 \geq 100$, made the basic solution corresponding to $x_1 = x_2 = 0$ infeasible.

To see this another way look at equations (3.5) and (3.6) in the Standard form representation of the problem. When x_1 and x_2 are set equal to zero, (3.5) and (3.6) reduce to

$$- 1s_5 = 100$$

and

$$- 1s_6 = 100.$$

Thus setting x_1 and x_2 equal to zero gives us the basic solution

$$
\begin{bmatrix}
s_1 \\
s_2 \\
s_3 \\
s_4 \\
s_5 \\
s_6
\end{bmatrix}
=
\begin{bmatrix}
630 \\
600 \\
708 \\
135 \\
-100 \\
-100
\end{bmatrix}
$$

Clearly this is not a basic feasible solution since s_5 and s_6 violate the nonnegativity requirements. Thus our former method of creating an initial basic feasible solution by setting each of the decision variables to zero will not work. The difficulty here is that Standard form and Tableau form are only equivalent for problems with less-than-or-equal-to constraints.

In order to set up Tableau form for this problem, we shall resort to a mathematical "trick" that will enable us to find an initial basic feasible solution in terms of the slack variables s_1, s_2, s_3, and s_4, and two new variables we shall denote as a_1 and a_2. These two new variables constitute the mathematical "trick". Variables a_1 and a_2 really have nothing to do with the original Par, Inc. problem, but merely serve to enable us to set up Tableau form and thus obtain an initial basic feasible solution. Since these new variables have been artificially created by us in order to just get things going we will refer to such variables as *artificial variables*. We caution the student to avoid confusing the notation for artificial variables with that used for elements of the A matrix. Elements of the A matrix always have two subscripts, whereas artificial variables only have one.

With the addition of two artificial variables, we can convert our Standard form representation of the modified Par, Inc. problem into Tableau form. We add an artificial variable, a_1, to equation (3.5) and an artificial variable, a_2, to equation (3.6) to obtain the following representation of our system of equations in Tableau form.

$$\frac{7}{10}x_1 + 1x_2 + 1s_1 \qquad\qquad\qquad\qquad\qquad = 630$$
$$\frac{1}{2}x_1 + \frac{5}{6}x_2 \qquad +1s_2 \qquad\qquad\qquad\qquad = 600$$
$$1x_1 + \frac{2}{3}x_2 \qquad\quad + 1s_3 \qquad\qquad\qquad = 708$$
$$\frac{1}{10}x_1 + \frac{1}{4}x_2 \qquad\qquad\quad +1s_4 \qquad\qquad = 135$$
$$1x_1 \qquad\qquad\qquad\qquad\qquad -1s_5 \quad +1a_1 \quad = 100$$
$$1x_2 \qquad\qquad\qquad\qquad\qquad\qquad -1s_6 \ +1a_2 = 100$$
$$x_1, x_2, s_1, s_2, s_3, s_4, s_5, s_6, a_1, a_2 \geq 0$$

Since the variables s_1, s_2, s_3, s_4, a_1, and a_2 each appear only once with a coefficient of "1", and since the right-hand sides are non-negative, both of the requirements of Tableau form have been satisfied.

We can now obtain an initial basic feasible solution to the system of equations in Tableau form by setting $x_1 = x_2 = s_5 = s_6 = 0$. This complete solution is

$$
\begin{bmatrix} x_1 \\ x_2 \\ s_1 \\ s_2 \\ s_3 \\ s_4 \\ s_5 \\ s_6 \\ a_1 \\ a_2 \end{bmatrix}
=
\begin{bmatrix} 0 \\ 0 \\ 630 \\ 600 \\ 708 \\ 135 \\ 0 \\ 0 \\ 100 \\ 100 \end{bmatrix}
$$

Is this solution feasible in terms of our real-world problem? No! It does not satisfy the requirements that we produce at least one hundred each of Standard and Deluxe bags. Thus we must make an important distinction between a basic feasible solution for the Tableau form of our problem and a basic feasible solution for the real-world problem. A basic feasible solution for the Tableau form of a linear programming problem is not always a basic feasible solution to the real-world problem. This is because of the appearance of the artificial variables in the Tableau form of the problem. However, since the Standard form representation of the problem does not include any of these artificial variables, a basic feasible solution for the Standard form representation will be feasible for the real-world problem. We see, then, that the Standard form representation is equivalent to the original problem whereas, whenever we have to add artificial variables, the Tableau form representation is not.

As we saw in Chapter 2, the reason for creating the Tableau form was to obtain an initial basic feasible solution to get the Simplex method started. Thus we see that whenever it is necessary to introduce artificial variables, the initial Simplex solution will not in general be feasible for the real-world problem. This situation is not as difficult as it might seem, however, since the only time we *must* have a feasible solution is at the *last* iteration of the Simplex method (i.e., the optimal solution must be feasible). Thus, if we could devise some means to guarantee that the artificial variables would be driven out of the basis before the optimal solution was reached there would be no difficulty.

The way in which we guarantee that these artificial variables will be driven out before the optimal solution is reached is to assign a very large cost to each of these variables in the objective function. For example, in the problem we are currently considering, we assign a very large negative number as the profit coefficient of each artificial variable in the objective function of the Tableau form. Hence, if these variables are in solution, they will necessarily be substantially reducing profits. As a result, these variables will be eliminated from the basis as soon as possible, and this is precisely what we want to happen.

As an alternative to picking a large negative number like $-100,000$ for the profit coefficient we will denote the profit coefficient of each artifical variable by $-M$. Here it is assumed that $-M$ represents some very large negative number. This notation will make it easier for us to keep track of the elements of the Simplex tableau which depend on the profit coefficients of the artificial variables. Using $-M$ as the profit coefficient for the artificial variables we can now write the objective function for the Tableau form of our problem.

$$\max \ z = 10x_1 + 9x_2 + 0s_1 + 0s_3 + 0s_4 + 0s_5 + 0s_6 - Ma_1 - Ma_2$$

Thus, in terms of our new artificial variables a_1 and a_2, we can now write the following initial Simplex tableau.

BASIS	c_j	x_1 10	x_2 9	s_1 0	s_2 0	s_3 0	s_4 0	s_5 0	s_6 0	a_1 $-M$	a_2 $-M$	
s_1	0	7/10	1	1	0	0	0	0	0	0	0	630
s_2	0	1/2	5/6	0	1	0	0	0	0	0	0	600
s_3	0	1	2/3	0	0	1	0	0	0	0	0	708
s_4	0	1/10	1/4	0	0	0	1	0	0	0	0	135
a_1	$-M$	①1	0	0	0	0	0	-1	0	1	0	100
a_2	$-M$	0	1	0	0	0	0	0	-1	0	1	100
	z_j	$-M$	$-M$	0	0	0	0	M	M	$-M$	$-M$	$-200M$
	$c_j - z_j$	$10+M$	$9+M$	0	0	0	0	$-M$	$-M$	0	0	

The above tableau corresponds to the solution $s_1 = 630$, $s_2 = 600$, $s_3 = 708$, $s_4 = 135$, $a_1 = 100$, $a_2 = 100$, and $x_1 = x_2 = s_5 = s_6 = 0$. In terms of our tableau, this is a basic feasible solution since all of the variables are ≥ 0 and $(n - m)$ of the variables are equal to zero. However, in terms of our modified Par, Inc. problem, $x_1 = x_2 = 0$ is clearly not feasible. This difficulty is caused by the fact that the artificial variables are in our current basic solution at positive values. Let us complete the Simplex solution to this problem and see if the artificial variables are driven out of solution as we hope they will be.

We see that at the first iteration x_1 will be brought into the basis and a_1 will be driven out. The Simplex tableau after this iteration is presented below.

Result of Iteration 1

BASIS	c_j	x_1 10	x_2 9	s_1 0	s_2 0	s_3 0	s_4 0	s_5 0	s_6 0	a_1 $-M$	a_2 $-M$	
s_1	0	0	1	1	0	0	0	7/10	0	$-7/10$	0	560
s_2	0	0	5/6	0	1	0	0	1/2	0	$-1/2$	0	550
s_3	0	0	2/3	0	0	1	0	1	0	-1	0	608
s_4	0	0	1/4	0	0	0	1	1/10	0	$-1/10$	0	125
x_1	10	1	0	0	0	0	0	-1	0	1	0	100
a_2	$-M$	0	①	0	0	0	0	0	-1	0	1	100
	z_j	10	$-M$	0	0	0	0	-10	M	10	$-M$	$1000-100M$
	c_j-z_j	0	$9+M$	0	0	0	0	10	$-M$	$-M-10$	0	

Note: The first 2 solutions are infeasible.

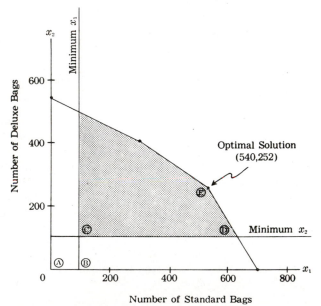

Figure 3–B.   Sequence of Simplex Solutions to Modified Par, Inc. Problem

The current solution is still not feasible since artificial variable a_2 is in the basis at a positive value. It does not satisfy the $x_2 \geq 100$ requirement. Graphically, we see in Figure 3–B above that this iteration has moved us from the origin (labelled Ⓐ in the figure) to point Ⓑ which is still not in the feasible region.

At the next iteration x_2 will be brought into solution and a_2 will be driven out. The Simplex tableau after this iteration is presented below.

Result of Iteration 2

BASIS	c_j	x_1 10	x_2 9	s_1 0	s_2 0	s_3 0	s_4 0	s_5 0	s_6 0	a_1 $-M$	a_2 $-M$	
s_1	0	0	0	1	0	0	0	7/10	1	$-7/10$	-1	460
s_2	0	0	0	0	1	0	0	1/2	5/6	$-1/2$	$-5/6$	466⅔
s_3	0	0	0	0	0	1	0	①	2/3	-1	$-2/3$	541⅓
s_4	0	0	0	0	0	0	1	1/10	1/4	$-1/10$	$-1/4$	100
x_1	10	1	0	0	0	0	0	-1	0	1	0	100
x_2	9	0	1	0	0	0	0	0	-1	0	1	100
z_j		10	9	0	0	0	0	-10	-9	10	9	1900
$c_j - z_j$		0	0	0	0	0	0	10	9	-10	-9	

The current solution is now feasible since all of the artificial variables have been driven out of solution. Thus, we now have the situation where the basic feasible solution contained in the Simplex tableau is also a basic feasible solution to the real-world problem. As you can see from Figure 3–B the current solution corresponds to point Ⓒ on the corner of the feasible region.

The next two iterations of the Simplex method move us from point Ⓒ to Ⓓ and from Ⓓ to Ⓔ on our graph. The resulting Simplex tableaus are given below.

Result of Iteration 3

BASIS	c_j	x_1 10	x_2 9	s_1 0	s_2 0	s_3 0	s_4 0	s_5 0	s_6 0	a_1 $-M$	a_2 $-M$	
s_1	0	0	0	1	0	$-7/10$	0	0	⟨16/30⟩	0	$-16/30$	2432/30
s_2	0	0	0	0	1	$-1/2$	0	0	3/6	0	$-3/6$	588/3
s_5	0	0	0	0	0	1	0	1	2/3	-1	$-2/3$	1624/3
s_4	0	0	0	0	0	$-1/10$	1	0	11/60	0	$-11/60$	1376/30
x_1	10	1	0	0	0	1	0	0	2/3	0	$-2/3$	1924/3
x_2	9	0	1	0	0	0	0	0	-1	0	1	100
z_j		10	9	0	0	10	0	0	$-7/3$	0	7/3	7313⅓
c_j-z_j		0	0	0	0	-10	0	0	7/3	$-M$	$-M-7/3$	

Result of Iteration 4

BASIS	c_j	x_1 10	x_2 9	s_1 0	s_2 0	s_3 0	s_4 0	s_5 0	s_6 0	a_1 $-M$	a_2 $-M$	
s_6	0	0	0	30/16	0	$-210/160$	0	0	1	0	-1	152
s_2	0	0	0	$-15/16$	1	25/160	0	0	0	0	0	120
s_5	0	0	0	$-20/16$	0	300/160	0	1	0	-1	0	440
s_4	0	0	0	$-11/32$	0	45/320	1	0	0	0	0	18
x_1	10	1	0	$-20/16$	0	300/160	0	0	0	0	0	540
x_2	9	0	1	30/16	0	$-210/160$	0	0	0	0	0	252
z_j		10	9	70/16	0	111/16	0	0	0	0	0	7668
c_j-z_j		0	0	$-70/16$	0	$-111/16$	0	0	0	$-M$	$-M$	

Just as with the graphical approach we see that the addition of the two greater-than-or-equal-to constraints has not changed our optimal solution. However, it has taken us more iterations to get to this point. This is because it took us two iterations to eliminate the artificial variables and hence obtain a basic feasible solution for the real-world problem.

Fortunately, once we obtain the initial Simplex tableau using artificial variables, we need not concern ourselves with worrying about whether the basic solution at a particular iteration is feasible for the real-world problem. We need only follow all the rules for the Simplex method. If we reach the stopping criterion (i. e., all $c_j - z_j \leq 0$) and all the artificial variables have been eliminated from solution, then we have found the optimal solution to our linear program. On the other hand, if we reach the stopping criterion and one or more of the artificial variables remains in solution at a positive value then there is no feasible solution to the real-world problem. This special case will be discussed in detail in Section 3.8.

3.3   Equality Constraints

When an equality constraint occurs in a linear programming problem, we need only add an artificial variable to get an initial basic feasible solution for our Simplex tableau. For example, if we had the equality constraint

$$6x_1 + 4x_2 - 5x_3 = 30,$$

we would simply add an artificial variable, say a_1, to enable us to create an initial basic feasible solution in our tableau. The above equation would then become

$$6x_1 + 4x_2 - 5x_3 + a_1 = 30.$$

Once we have created Tableau form by adding artificial variables to all of the equality constraints, the Simplex method proceeds exactly like the case for the greater-than-or-equal-to constraints situation.

3.4   Negative Right-Hand Sides

At this point, we have just about considered all cases that can arise with respect to the constraint equations of our linear programming problem. One more point, however, deserves men-tion. What if one or more of the values on the right-hand side of our constraints are *negative*? For example, suppose that management of Par, Inc. had specified that the number of Standard bags produced had to be less than or equal to the number of Deluxe bags after 25 Deluxe bags had been saved for display purposes. We could formulate this constraint as

$$x_1 \leq x_2 - 25.$$

Subtracting x_2 from both sides of the inequality allows us to place all the variables on the left-hand side of the constraint and the constant on the right-hand side. Thus we have

$$x_1 - x_2 \leq - 25.$$

Our standard procedure of adding a slack variable ($x_1 - x_2 + s_1 = - 25$) to obtain Tableau form is unacceptable since our constraint would not satisfy the Tableau form requirement of non-negative right-hand sides. Thus, we must look for ways to remove the negative right-hand side values before we can set up our Tableau form representation of the problem. As you will see, this is relatively easy to do. There are three separate cases to consider. That is, we must consider whether the constraint in question is an equality, greater-than-or-equal-to, or less-than-or-equal-to constraint.

Case (1) Equality constraint—For example,

$$6x_1 + 3x_2 - 4x_3 = - 20.$$

We need only multiply both sides of the equation by -1 and then introduce the artificial variable. Hence, we first get

$$- 6x_1 - 3x_2 + 4x_3 = 20,$$
and then
$$- 6x_1 - 3x_2 + 4x_3 + a_1 = 20.$$

Thus, our initial basic feasible solution will show $a_1 = 20$.

Case (2) Greater-than-or-equal-to constraint—For example,

$$6x_1 + 3x_2 - 4x_3 \geq - 20.$$

What would happen if we multiplied both sides by -1? The rule is that if you multiply both sides of an inequality by a negative number, the sign of the inequality changes direction. For example, the inequality $1 \geq -2$ is certainly true. However, if we multiply both sides by -1 we must change the direction of the inequality in order to have the correct relationship $-1 \leq 2$. Similarly, multiplying the above constraint by -1 and changing the direction of the inequality yields

$$-6x_1 - 3x_2 + 4x_3 \leq 20.$$

This constraint can now be treated the same as any ordinary less-than-or-equal-to constraint by adding a slack variable to the left-hand side and making the slack variable part of the initial basic feasible solution.

Case (3) Less-than-or-equal-to constraint—For example,

$$6x_1 + 3x_2 - 4x_3 \leq -20.$$

We multiply both sides by -1 and change the direction of the inequality to get $-6x_1 - 3x_2 + 4x_3 \geq 20$.

Now we have the usual situation for a greater-than-or-equal-to constraint. That is, all we need to do now to obtain Tableau form is to change the left-hand side by subtracting a surplus variable and adding an artificial variable. The artificial variable will then appear in the initial basic feasible solution.

Summarizing, we see that any time the original formulation of a linear program contains a negative right-hand side we must perform the preliminary operations outlined above before adding slack, artificial and surplus variables.

3.5 Summary of the Constraint Situation

Let us pause for a moment and review the procedures we have established for handling constraints in preparation for solving a linear program by the Simplex method. These are the general procedures to be used to obtain Tableau form.

1. For $\leq$ constraints we simply add a slack variable to each less-than-or-equal-to constraint to obtain an equality. The coefficient of the slack variable in the objective function is assigned a value of zero. This gives us Tableau form and the slack variable becomes one of the variables in our initial basic feasible solution.

2. For equality constraints, we add an artificial variable to each equality constraint to obtain the Tableau form. The coefficient of this artificial variable in the objective function is assigned a value of $-M$. This artificial variable becomes part of our initial basic feasible solution.

3. For $\geq$ constraints we subtract a surplus variable to obtain an equality. Then, we add an artificial variable to get the Tableau form of our problem. This artificial variable becomes part of our initial basic feasible solution. The coefficient of the surplus variable in the objective function is zero and the coefficient of the artificial variable is $-M$.

To get some practice applying the above principles, let us now convert the following numerical example into Tableau form and set up the initial Simplex tableau.

$$\max 6x_1 + 3x_2 + 4x_3 + x_4$$

s.t.

$$2x_1 + \tfrac{1}{2}x_2 - x_3 + 6x_4 = 60$$
$$x_1 \quad\quad + x_3 + \tfrac{2}{3}x_4 \leq 20$$
$$x_2 + 5x_3 \quad\quad \geq 50$$
$$x_1, x_2, x_3, x_4 \geq 0.$$

Recall that in order to obtain Standard form, all of the constraints must be made equalities. Thus, by using slack and surplus variables where appropriate, we obtain the following Standard form representation.

$$\max 6x_1 + 3x_2 + 4x_3 + x_4 + 0s_1 + 0s_2$$

s.t.

$$2x_1 + \tfrac{1}{2}x_2 - 1x_3 + 6x_4 \quad\quad\quad\quad = 60 \quad\quad (3.7)$$
$$1x_1 + 0x_2 + 1x_3 + \tfrac{2}{3}x_4 + 1s_1 \quad\quad = 20 \quad\quad (3.8)$$
$$0x_1 + 1x_2 + 5x_3 + 0x_4 \quad\quad - 1s_2 = 50 \quad\quad (3.9)$$
$$x_1, x_2, x_3, x_4, s_1, s_2, \geq 0$$

In order to obtain Tableau form, we must add an artificial variable to equations (3.7) and (3.9). Adding artificial variable a_1

to equation (3.7) and artificial variable a_2 to equation (3.9), we get

$$\max \; 6x_1 + 3x_2 + 4x_3 + \; x_4 + 0s_1 + 0s_2 - Ma_1 - Ma_2$$

s.t.

$$2x_1 + \tfrac{1}{2}x_2 - 1x_3 + 6x_4 \qquad\qquad + 1a_1 \qquad\qquad = 60$$
$$1x_1 + 0x_2 + 1x_3 + \tfrac{2}{3}x_4 + 1s_1 \qquad\qquad\qquad\qquad = 20$$
$$0x_1 + 1x_2 + 5x_3 + 0x_4 \qquad - 1s_2 \qquad + 1a_2 = 50$$

$$x_1, x_2, x_3, x_4, s_1, s_2, a_1, a_2 \geq 0 .$$

The initial Simplex tableau corresponding to this Tableau form is

BASIS	c_j	x_1	x_2	x_3	x_4	s_1	s_2	a_1	a_2	
		6	3	4	1	0	0	$-M$	$-M$	
a_1	$-M$	2	1/2	-1	6	0	0	1	0	60
s_1	0	1	0	1	2/3	1	0	0	0	20
a_2	$-M$	0	1	5	0	0	-1	0	1	50
	z_j	$-2M$	$-1\tfrac{1}{2}M$	$-4M$	$-6M$	0	M	$-M$	$-M$	$-110M$
	$c_j - z_j$	$2M+6$	$1\tfrac{1}{2}M+3$	$4M+4$	$6M+1$	0	$-M$	0	0	

3.6 The Minimization Problem

Many linear programming problems are more naturally formulated as minimization problems. For example, a manufacturer who has contracted to sell a certain number of units of his product to various buyers. He is no longer concerned with how many units to produce. His problem is one of minimizing the total cost of production subject to the constraints that he must satisfy demand. As an illustration of how this type of minimization problem might occur, consider the problem encountered by Photo Chemicals, Inc.

Photo Chemicals produces two types of picture-developing fluids. Both products cost Photo Chemicals $1.00 per gallon to produce. Based upon an analysis of current inventory levels and outstanding orders for the next month, Photo's management has specified that at least 30 gallons of product 1 and at least 20 gallons of product 2 must be produced during the next two weeks. In addition, management has stated that an existing inventory of highly perishable raw material that is used in the production of both fluids must be used within the next two weeks. Other than these requirements management, because of some serious working capital limitations, would like to keep production costs at the minimum possible level over the next two week period.

The current inventory of the perishable raw material is 80 pounds. While more of this raw material can be ordered if necessary, any that is not used within the next two weeks will spoil; hence, the management requirement that at least 80 pounds must be used in the next two weeks. Furthermore, it is known that product 1 requires one pound of this perishable raw material per gallon produced, and product 2 requires two pounds of the raw material per gallon.

In summary, management is looking for a minimum cost production plan that uses at least the 80 pounds of perishable raw material currently in inventory, and provides at least 30 gallons of product 1 and 20 gallons of product 2. What is this minimum cost solution? Consider the following linear programming formulation of our problem, where

$$x_1 = \text{number of gallons of product 1 produced}$$

and

$$x_2 = \text{number of gallons of product 2 produced.}$$

$$\min x_1 + x_2$$

s.t.

$$x_1 \qquad \geq 30 \quad (\text{product 1})$$
$$x_2 \geq 20 \quad (\text{product 2})$$
$$x_1 + 2x_2 \geq 80 \quad (\text{raw material})$$
$$x_1, x_2 \geq 0$$

In Figure 3–C we show the constraint equations and the feasible region corresponding to our problem.

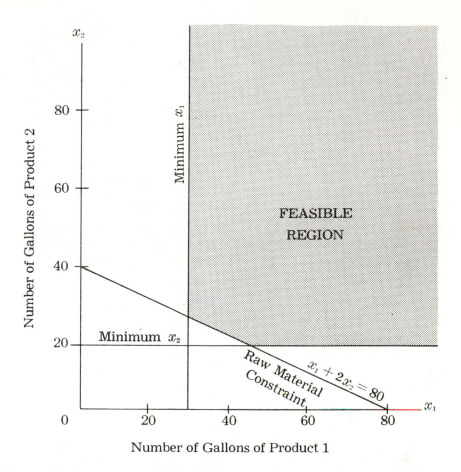

Figure 3–C.   Set of Feasible Solutions for Photo Chemicals, Inc.

In order to determine the minimum value of our cost, $(x_1 + x_2)$, we begin by drawing the cost line corresponding to a particular value of cost, $z = x_1 + x_2$. For example, we might start by drawing the line $x_1 + x_2 = 80$. In Figure 3–D we show the equation of this line. Clearly, there are many points in the feasible region yielding this cost value (e. g., $x_1 = x_2 = 40$).

To find the values of x_1 and x_2 which yield the optimal solution we move our cost line in a lower-left direction until, if we moved it any further, it would be entirely outside the feasible region. We see that the line $x_1 + x_2 = 55$ intersects the feasible region at

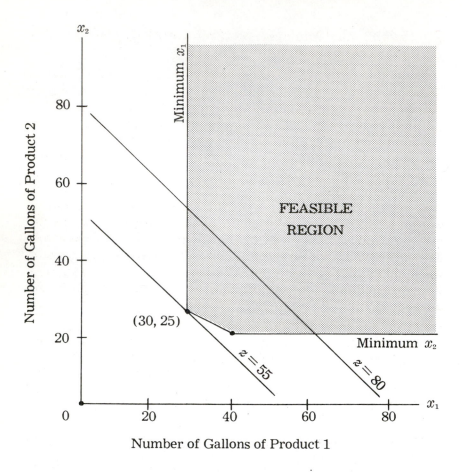

Figure 3–D.   Graphical Solution for Photo Chemicals, Inc.
              Problem

the point ($x_1 = 30$, $x_2 = 25$) and that if we move the cost line any
lower it will be outside of the feasible region.   Thus the optimal
solution to our problem is $x_1 = 30$, $x_2 = 25$, with a corresponding
objective function value of 55.   Note in Figure 3–D that the raw
material constraint and the minimum x_1 constraints are binding.

3.7   Solving the Minimization Problem Using the Simplex Method

There are two ways in which we can solve a minimization problem using the Simplex method. The first requires that we change the rule used to introduce a variable into solution. Recall that in the maximization case, we selected the variable with the largest positive (c_j-z_j) as the variable to introduce next into the basis. This was because the value of (c_j-z_j) told us the amount the objective function would increase if one unit of the variable in column j was brought into the basis. To solve the minimization problem we can simply reverse this rule. That is, we can select the variable with the most negative (c_j-z_j) as the one to introduce next. Of course, this means our stopping rule will also have to be changed. In the minimization case, we stop when every value in the net evaluation row is non-negative. When this condition occurs, we have an optimal solution to the minimization problem.

Let us now look at the second way in which we can solve a minimization problem using the Simplex method. This second approach is the one we shall use in the remainder of the book whenever it is required to solve a minimization problem. The approach relys on a well-known mathematical "trick" often employed in optimization problems. It turns out that if one wishes to solve the problem, min c^tx subject to a set of constraints (linear or otherwise), an equivalent problem is max $-c^tx$ subject to the same constraints. These problems are equivalent in the sense that the same value of the vector x which solves min c^tx also solves max $-c^tx$. The only difference is that the value of the solution to max $-c^tx$ is the negative of the solution to min c^tx. That is,

$$\min c^tx = - \max - c^tx.$$

Consider the data in Table 3–A. which shows the values of c^tx and $-c^tx$ for selected feasible solutions to our Photo Chemicals, Inc. problem. As you can see, the values of x_1 and x_2 that minimize c^tx are also the values of x_1 and x_2 that maximize $-c^tx$. Moreover, we see that the value of the solution that minimizes $c^tx = x_1 + x_2$, i. e., $z = 55$, is the negative of the value of the solution that maximizes $-c^tx = - x_1 - x_2$. Thus, we see that if we want to solve min $(x_1 + x_2)$, we need only solve the problem max $(-x_1 - x_2)$ and multiply the value of the solution to max $(-x_1 - x_2)$ by -1. This relationship will always hold true, and will be the method we shall always use to solve minimization problems.

Selected Feasible Solutions		$c^t x = x_1 + x_2$	$-c^t x = -x_1 - x_2$
$x_1 = 40$	$x_2 = 40$	80	-80
$x_1 = 40$	$x_2 = 30$	70	-70
$x_1 = 40$	$x_2 = 20$	60	-60
$x_1 = 30$	$x_2 = 40$	70	-70
$x_1 = 30$	$x_2 = 30$	60	-60
$x_1 = 30$	$x_2 = 25$	55 (min value of $c^t x$)	-55 (max value of $-c^t x$)

Table 3–A. A Comparison of Feasible Solutions to Show the Min $c^t x$ Solution is the Max $- c^t x$ Solution.

Employing the max $-c^t x$ approach to solving the minimization problem means that we can follow exactly the same Simplex solution procedure that was outlined for the maximization problem earlier. The only change necessary is in setting up the Standard form representation of our problem. That is, we multiply the objective function by -1 before setting up the Standard form representation. Let us see how this procedure works for the Photo Chemicals problem we just solved graphically.

We previously saw that the Photo Chemicals problem could be formulated as

$$\min \; x_1 + \; x_2$$

s.t.

$$x_1 \qquad\;\; \geq 30$$
$$x_2 \geq 20$$
$$x_1 + 2x_2 \geq 80$$
$$x_1, x_2 \geq 0 \;.$$

To solve the problem using our maximization Simplex procedure, we first multiply the objective function by a -1 in order to convert our minimization problem into the following equivalent maximization problem.

max $-x_1 - x_2$

s.t.

$$x_1 \qquad\qquad \geq 30$$
$$x_2 \geq 20$$
$$x_1 + 2x_2 \geq 80$$
$$x_1, x_2 \geq 0$$

After subtracting surplus variables, we obtain the following Standard form representation for our problem.

max $- x_1 - x_2 + 0s_1 + 0s_2 + 0s_3$

s.t.

$$1x_1 + 0x_2 - 1s_1 \qquad\qquad = 30$$
$$0x_1 + 1x_2 \qquad -1s_2 \qquad = 20$$
$$1x_1 + 2x_2 \qquad\qquad -1s_3 = 80$$
$$x_1, x_2, s_1, s_2, s_3 \geq 0$$

Since our problem is one involving $\geq$ constraints, we must add artificial variables to obtain Tableau form. After adding artificial variables to each of our constraints we get the following Tableau form for our Photo Chemicals problem.

max $-x_1 - x_2 + 0s_1 + 0s_2 + 0s_3 - Ma_1 - Ma_2 - Ma_3$

s.t.

$$1x_1 + 0x_2 - 1s_1 \qquad\qquad + 1a_1 \qquad\qquad = 30$$
$$0x_1 + 1x_2 \qquad -1s_2 \qquad\qquad + 1a_2 \qquad = 20$$
$$1x_1 + 2x_2 \qquad\qquad - 1s_3 \qquad\qquad + 1a_3 = 80$$
$$x_1, x_2, s_1, s_2, s_3, a_1, a_2, a_3, \geq 0$$

The initial Simplex tableau becomes:

BASIS	c_j	x_1	x_2	s_1	s_2	s_3	a_1	a_2	a_3	
		-1	-1	0	0	0	$-M$	$-M$	$-M$	
a_1	$-M$	1	0	-1	0	0	1	0	0	30
a_2	$-M$	0	1	0	-1	0	0	1	0	20
a_3	$-M$	1	2	0	0	-1	0	0	1	80
	z_j	$-2M$	$-3M$	M	M	M	$-M$	$-M$	$-M$	$-130M$
	c_j-z_j	$-1+2M$	$-1+3M$	$-M$	$-M$	$-M$	0	0	0	

Let us now work through the solution to this problem.

First Iteration. In our example, x_2 is the variable with the most positive $(c_j - z_j)$; thus, we choose variable x_2 as the variable to introduce into solution. To determine which variable leaves the basis, we perform the usual calculations. Calculating the ratios $\dfrac{\bar{b}_i}{\bar{a}_{i2}}$ for $\bar{a}_{i2} > 0$, we see that $20/1$ is the smallest positive ratio, and corresponds to basic variable a_2. What does this tell us? Well, it says that if we introduce 20 units of x_2 into the basis, we will have to set $a_2 = 0$ in order to satisfy the second constraint. Hence, a_2 will be driven out of solution when x_2 is introduced at a level of 20 units. We have circled the pivot element, $\bar{a}_{22}$, in the tableau. After performing the usual row operations to make x_2 a basic variable, we can write the second tableau as follows:

BASIS	c_j	x_1	x_2	s_1	s_2	s_3	a_1	a_2	a_3	
		-1	-1	0	0	0	$-M$	$-M$	$-M$	
a_1	$-M$	①	0	-1	0	0	1	0	0	30
x_2	-1	0	1	0	-1	0	0	1	0	20
a_3	$-M$	1	0	0	2	-1	0	-2	1	40
	z_j	$-2M$	-1	M	$-2M+1$	M	$-M$	$-1+2M$	$-M$	$-70M-20$
	$c_j - z_j$	$1+2M$	0	$-M$	$-1+2M$	$-M$	0	$-3M+1$	0	

Second Iteration. Since x_1 and s_2 are tied for the largest positive $(c_j - z_j)$, it makes no difference which one we introduce. Just in order to have a procedure to follow in the case of ties, we will follow the convention of introducing the variable on the left whenever there is a tie. Calculating the ratios $\dfrac{\bar{b}_i}{\bar{a}_{i1}}$ for $\bar{a}_{i1} > 0$, we see that $30/1$, corresponding to a_1, is the smallest ratio. Therefore, a_1 will be driven out of solution when x_1 is introduced. The pivot element is again circled. After performing the necessary pivot operations, we can write the third tableau.

BASIS	c_j	x_1	x_2	s_1	s_2	s_3	a_1	a_2	a_3	
		-1	-1	0	0	0	$-M$	$-M$	$-M$	
x_1	-1	1	0	-1	0	0	1	0	0	30
x_2	-1	0	1	0	-1	0	0	1	0	20
a_3	$-M$	0	0	1	(2)	-1	-1	-2	1	10
z_j		-1	-1	$-M+1$	$-2M+1$	M	$-1+M$	$-1+2M$	$-M$	$-10M-50$
c_j-z_j		0	0	$-1+M$	$-1+2M$	$-M$	$-2M+1$	$-3M+1$	0	

Third Iteration. Here, our criterion for entering a variable indicates that s_2 is to be introduced into solution. Notice that introduction of s_2 will drive the final artificial variable, a_3 out of solution. Hence, after this iteration, we will have a feasible solution to our real-world problem. Furthermore, if $(c_j - z_j)$ ≤ 0 for all the variables, this iteration will yield the optimal solution. Let us calculate the next tableau and see what results.

BASIS	c_j	x_1	x_2	s_1	s_2	s_3	a_1	a_2	a_3	
		-1	-1	0	0	0	$-M$	$-M$	$-M$	
x_1	-1	1	0	-1	0	0	1	0	0	30
x_2	-1	0	1	$1/2$	0	$-1/2$	$-1/2$	0	$1/2$	25
s_2	0	0	0	$1/2$	1	$-1/2$	$-1/2$	-1	$1/2$	5
z_j		-1	-1	$1/2$	0	$1/2$	$1/2$	0	$-1/2$	-55
c_j-z_j		0	0	$-1/2$	0	$-1/2$	$-M-1/2$	$-M$	$-M+1/2$	

AHA! It turns out that the third iteration does provide us with the optimal solution (all $c_j - z_j$ values are ≤ 0). Looking back to the solution obtained using the graphical procedure, we see that this is indeed the same solution.

Now, look back at Figure 3–C and look at the path we followed in going from the origin to the optimal solution. We started at the origin ($x_1 = 0$, $x_2 = 0$) with our initial Simplex tableau. Our first iteration took us from the origin to the point ($x_1 = 0$, $x_2 = 20$). Note that at this point we are still in the infeasible region. The second iteration took us to the point ($x_1 = 30$, $x_2 = 20$) as x_1 was introduced into solution. However we still did not have a feasible solution to our real-world problem. Finally, the last iteration took us to ($x_1 = 30$, $x_2 = 25$). This is a feasible solution as we can easily verify from the graph. Indeed, it is also the optimal solution to our problem. Note that the optimal solution from our tableau shows that we produce 30 units of product 1 and 25 units of product 2 and that with $s_2 = 5$, we will have a surplus of 5 units of product 2 over what was required.

We now have the ability to solve minimization problems as well as maximization problems. Actually we had this ability all along. It's just that we did not recognize until now that any minimization problem could be converted to an equivalent maximization problem by simply multiplying the objective function by (-1). We shall now concentrate on discussing some important special cases that may occur when we are trying to solve any linear programming problem. In the following discussion we will only consider the case for maximization problems, recognizing that all minimization problems may be placed in this form.

3.8   Infeasibility

Infeasibility comes about when there is no solution to the linear programming problem which satisfies all the constraints, including the nonnegativity conditions, x_1, x_2, . . . , $x_n \geq 0$. Graphically, infeasibility means that a feasible region does not exist. That is, there are no points which satisfy all the constraining equations and the nonnegativity constraints simultaneously. To illustrate this situation let us look again at the problem faced by Par, Inc.

Suppose that instead of specifying that 100 of each bag be manufactured, management had specified that at least 500 of the Standard bags and 360 of the Deluxe bags must be manufactured. The graph of our solution region may now be constructed to reflect these requirements.

The shaded area in the lower left-hand portion of the graph depicts those points satisfying the departmental constraints on the availability of time. The shaded area in the upper right-hand portion depicts those points satisfying the minimum pro-

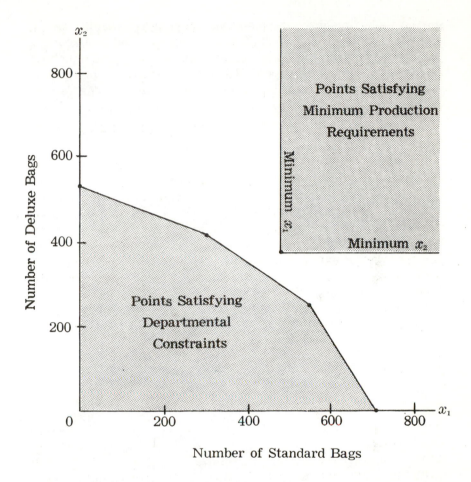

Figure 3–E. No Feasible Region for Par, Inc. Problem with Production Requirements of 500 Standard and 360 Deluxe Bags

duction requirements of 500 Standard and 360 Deluxe bags. But, there are no points satisfying both sets of constraints. Thus we see that if management imposes these minimum production requirements there will be no feasible solution to the linear programming model.

How should we interpret this infeasibility in terms of our current problem? Well, we should first tell management that, given the resources available (i.e., cutting and dyeing time, sewing time, finishing time, and inspection and packaging time), it is not possible to make 500 Standard bags and 360 Deluxe bags. Moreover, we can tell management specifically how much of each resource must be expended in order to make it possible to manufacture 360 Deluxe and 500 Standard bags. The following minimum amounts of resources must be available.

	Minimum Required Resources	Available Resources
Cutting & Dyeing:	$\frac{7}{10}(500) + 1(360) = 710$	630
Sewing:	$\frac{1}{2}(500) + \frac{5}{6}(360) = 550$	600
Finishing:	$1(500) + \frac{2}{3}(360) = 740$	708
Inspection & Packaging:	$\frac{1}{10}(500) + \frac{1}{4}(360) = 140$	135

Thus, we need 80 more hours of cutting and dyeing time, 32 more hours of finishing time, and 5 more hours of inspection and packaging time in order to meet management's requirement.

If after seeing our report, management still wants to manufacture 500 Standard and 360 Deluxe bags, it must somehow provide additional resources. Perhaps this will mean hiring another man to work in the cutting and dyeing department, transferring a person from elsewhere in the plant to work part-time in the finishing department, or having the sewing people help out periodically with the inspection and packaging. As you can see, there are many possibilities for corrective management action once we discover there is no feasible solution. The important thing for us to realize is that linear programming analysis can help us determine whether or not management's plans are feasible. By analyzing the problem using linear programming, we are often able to point out infeasible conditions and initiate corrective action.

Now that we have explored the implications of infeasibility in the context of a graphical solution, let us see how infeasibility is identified in the Simplex tableau. We mentioned earlier, in conjunction with artificial variables, that infeasibility in terms of the Simplex tableau meant that our stopping criteria would indicate an optimal solution while one or more of the artificial variables still remained in solution at a positive value. As an illustration of this phenomenon, we present the Simplex solution to the current modification of the Par, Inc. problem which we just solved graphically.

Initial Tableau

BASIS	c_j	x_1 10	x_2 9	s_1 0	s_2 0	s_3 0	s_4 0	s_5 0	s_6 0	a_1 $-M$	a_2 $-M$	
s_1	0	7/10	1	1	0	0	0	0	0	0	0	630
s_2	0	1/2	5/6	0	1	0	0	0	0	0	0	600
s_3	0	1	2/3	0	0	1	0	0	0	0	0	708
s_4	0	1/10	1/4	0	0	0	1	0	0	0	0	135
a_1	$-M$	①	0	0	0	0	0	-1	0	1	0	500
a_2	$-M$	0	1	0	0	0	0	0	-1	0	1	360
	z_j	$-M$	$-M$	0	0	0	0	M	M	$-M$	$-M$	$-860M$
	c_j-z_j	$M+10$	$M+9$	0	0	0	0	$-M$	$-M$	0	0	

Second Tableau

BASIS	c_j	x_1 10	x_2 9	s_1 0	s_2 0	s_3 0	s_4 0	s_5 0	s_6 0	a_1 $-M$	a_2 $-M$	
s_1	0	0	①	1	0	0	0	7/10	0	$-7/10$	0	280
s_2	0	0	5/6	0	1	0	0	1/2	0	$-1/2$	0	350
s_3	0	0	2/3	0	0	1	0	1	0	-1	0	208
s_4	0	0	1/4	0	0	0	1	1/10	0	$-1/10$	0	85
x_1	10	1	0	0	0	0	0	-1	0	1	0	500
a_2	$-M$	0	1	0	0	0	0	0	-1	0	1	360
	z_j	10	$-M$	0	0	0	0	-10	M	10	$-M$	$5000-360M$
	c_j-z_j	0	$M+9$	0	0	0	0	10	$-M$	$-M-10$	0	

Final Tableau

BASIS	c_j	x_1 10	x_2 9	s_1 0	s_2 0	s_3 0	s_4 0	s_5 0	s_6 0	a_1 $-M$	a_2 $-M$	
x_2	9	0	1	1	0	0	0	7/10	0	$-7/10$	0	280
s_2	0	0	0	$-5/6$	1	0	0	$-1/12$	0	1/12	0	116⅔
s_3	0	0	0	$-2/3$	0	1	0	16/30	0	$-16/30$	0	21⅓
s_4	0	0	0	1/4	0	0	1	$-9/120$	0	9/120	0	15
x_1	10	1	0	0	0	0	0	-1	0	1	0	500
a_2	$-M$	0	0	-1	0	0	0	$-7/10$	-1	7/10	1	80
z_j		10	9	$9+M$	0	0	0	$\dfrac{-37+7M}{10}$	M	$\dfrac{37-7M}{10}$	$-M$	$7520-80M$
c_j-z_j		0	0	$-9-M$	0	0	0	$\dfrac{37-7M}{10}$	$-M$	$\dfrac{-37-3M}{10}$	0	

Just as you might have expected, one of the artificial variables, a_2, is in the final solution. Notice that $(c_j - z_j) \leq 0$ for all the variables; therefore, according to the rules we established earlier, this should be the optimal solution. But, this solution is not feasible for our real-world problem since it has $x_1 = 500$ and $x_2 = 280$ (recall that we had to make at least 360 Deluxe bags). The fact that artificial variable a_2 is in solution at a value of 80 tells us that the final solution violates the sixth constraint ($x_2 \geq 360$) by 80 units.

If we are interested in knowing which constraints are preventing us from getting a feasible solution we can obtain at least a partial answer to this from our final Simplex tableau. Notice that $s_2 = 116⅔$, $s_3 = 21⅓$, and $s_4 = 15$. Since s_1 is not in solution, it has a value of zero. This tells us that the current solution uses all the cutting and dyeing time available but does not use 116⅔ hours of sewing time, 21⅓ hours of finishing time, and 15 hours of inspection and packaging time. Thus, what has actually happened is that the cutting and dyeing operation is causing a bottleneck. Since there is not enough cutting and dyeing time available, we cannot obtain the necessary x_1 and x_2 volumes. This occurs even though idle time exists in other departments.

The management implications here are that additional cutting and dyeing time should be made available in order to eliminate the bottleneck. After eliminating the problem in the cutting and dyeing department, it may still turn out that we cannot obtain a feasible solution. (This will obviously be the case for Par, Inc. since not enough finishing or inspection and packaging time is available to make the required number of bags.) Unless management decides to relax the requirement that 500 Standard bags and 360 Deluxe bags be manufactured, it will have to continue allocating resources to bottleneck departments until the linear programming problem has a feasible solution.

In summary, a linear program is infeasible if there is no solution which satisfies all of the constraints and nonnegativity conditions simultaneously. Graphically, we recognize this situation as the case where there is no feasible region. In terms of the Simplex solution procedure, we know that if one or more of the artificial variables remains in the final solution at a positive value there is no feasible solution to the real-world problem. In closing we note that for linear programming problems with $\leq$ constraints and nonnegative right-hand sides there will always be a feasible solution; since it is not necessary to introduce artificial variables to set up the initial Simplex tableau, there could not possibly be an artificial variable in the final solution.

3.9 Unboundedness

A solution to a linear programming problem is unbounded if the value of the solution may be made infinitely large without violating any of the constraints. This condition might be termed, "managerial utopia". If this condition were to occur in a profit maximization problem it would be true that the manager could achieve any level of profit he wanted.

In linear programming models of real-world problems the occurrence of an unbounded solution means the problem has been improperly formulated. That is, we know from our own experience and observations, that it is not possible to increase profits indefinitely. Therefore, we must conclude that if a profit maximization problem results in an unbounded solution, then improper problem formulation has occurred. That is, our mathematical model is not a sufficiently accurate representation of the real-world problem.

Graphically speaking, if a linear programming problem has an unbounded solution, the feasible region extends to infinity in some direction. As an illustration, consider the simple numerical example

$$\max\ 2x_1 + x_2$$

s.t.

$$x_1 \qquad \geq 2$$
$$x_2 \leq 5$$
$$x_1,\ x_2 \geq 0.$$

In Figure 3–F we have graphed the feasible region associated with this problem.

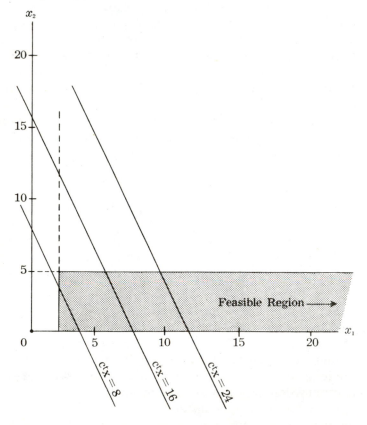

Figure 3–F.   Example of an Unbounded Problem

Note that we can only indicate part of the feasible region since the feasible region extends indefinitely in the direction of the x_1 axis. Looking at the profit lines in Figure 3–F, we see that the solution to this problem may be made as large as we desire. That is, no matter what solution we pick, there will always be some feasible solution with a larger value. Thus, we say that the solution to this linear program is unbounded.

If we are using the Simplex method of solution and the linear program is unbounded, we will automatically discover this before reaching the final tableau. What will happen is that the rule for determining the variable to remove from solution will not work. Recall that we calculated the ratio $\dfrac{\overline{b}_i}{\overline{a}_{ij}}$ for each of the elements of column j which were *positive*. Then we picked the smallest ratio to tell us which variable to remove from the current basic feasible solution.

As you will recall from Chapter 2, the coefficients in a particular column of $\overline{A}$ indicate how much each of the current basic variables will decrease if one unit of the variable associated with that particular column is brought into solution. For example, if $\overline{a}_{34} = 2$ then the value of the basic variable associated with the third row will decrease by 2 units if one unit of the variable associated with the fourth column is brought into solution. Suppose, then, that for a particular linear program we found that $(c_2 - z_2) = 5 > 0$, and that all the $\overline{a}_{i2}$ in column 2 were ≤ 0. This would mean that each unit of x_2 brought into solution would increase the objective function by five units. Furthermore, since $\overline{a}_{i2} \le 0$ for all i, this would mean that none of the current basic variables would be driven to zero no matter how many units of x_2 we introduced. Thus, we could introduce an infinite amount of x_2 into solution and still maintain feasibility. Since each unit of x_2 increases the objective function by five you can see that we would have an unbounded solution in this case. Hence, the way we recognize the unbounded situation is if all the $\overline{a}_{ij}$ are ≤ 0 in column j and the Simplex method indicates that variable x_j is to be introduced into solution.

To show this explicitly, let us solve our unbounded example problem using the Simplex method and see what happens. We first subtract a surplus variable, s_1, from the first constraint equation, and add a slack variable, s_2, to the second constraint equation to obtain Standard form. We then add an artificial variable, a_1, to the first constraint equation in order to obtain Tableau form and set up the initial Simplex tableau in terms of the basic variables a_1 and s_2. After bringing in x_1 at the first iteration, our Simplex tableau becomes

BASIS	c_j	x_1	x_2	s_1	a_1	s_2	
		2	1	0	$-M$	0	
x_1	2	1	0	-1	1	0	2
s_2	0	0	1	0	0	1	5
	z_j	2	0	-2	2	0	4
	$c_j - z_j$	0	1	2	$-M-2$	0	

Since s_1 has the largest positive $(c_j - z_j)$, we know we can increase the value of the objective function most rapidly by bringing s_1 into the basis. But $\bar{a}_{13} = -1$ and $\bar{a}_{23} = 0$; hence we cannot form the ratio $\dfrac{\bar{b}_i}{\bar{a}_{i3}}$ for all positive $\bar{a}_{i3}$, since there are none. This is our indication that the solution to the linear program is unbounded. We can interpret this condition as follows.

Each unit of s_1 that we bring into the basis drives 0 units of s_2 out of solution, and "gives" us an extra unit of x_1 since $\bar{a}_{13} = -1$. This is because s_1 is a surplus variable and can be interpreted as the amount of product 1 we produce over the minimum amount required, i.e., $x_1 = 2$. Since our Simplex tableau has indicated that we can introduce as much of s_1 as we desire without violating any constraints, this tells us that we can make as much as we want above the minimum amount of x_1 required. Thus, there will be no upper bound on the value of the objective function since the objective function coefficient associated with x_1 is positive.

In summary, a maximization linear program is unbounded if it is possible to make the value of the optimal solution as large as desired without violating any of the constraints. We can recognize this condition graphically as the case where the feasible region extends to infinity in some direction. When employing the Simplex solution procedure, an unbounded linear program is easy to recognize. That is, if at some iteration, the Simplex method tells us to introduce x_j into solution and all of the $\overline{a}_{ij}$ are less than or equal to zero in the jth column, we recognize that we have a linear program having an unbounded solution.

We emphasize that the case of an unbounded solution will never occur in real-world cost minimization or profit maximization problems because of the fact that it is not possible to reduce costs to minus infinity or to increase profits to plus infinity. Thus if we encounter this situation when solving a linear programming model in practice, we should go back and examine carefully our formulation of the problem to determine if we have made some error, or if the linear programming model is inappropriate.

3.10 Alternate Optimal Solutions

When we have a linear program which has two or more optimal solutions, we say the program has alternate optima. Graphically, this is the case where the objective function is parallel to one of the binding constraints. As an example, let us consider the original Par, Inc. problem with a slightly modified objective function.

$$\max\ 7x_1 + 10x_2$$

s.t.

$$\tfrac{7}{10}x_1 + 1x_2 \le 630$$

$$\tfrac{1}{2}x_1 + \tfrac{5}{6}x_2 \le 600$$

$$1x_1 + \tfrac{2}{3}x_2 \le 708$$

$$\tfrac{1}{10}x_1 + \tfrac{1}{4}x_2 \le 135$$

$$x_1,\ x_2 \ge 0$$

The graphical solution is presented in Figure 3–G.

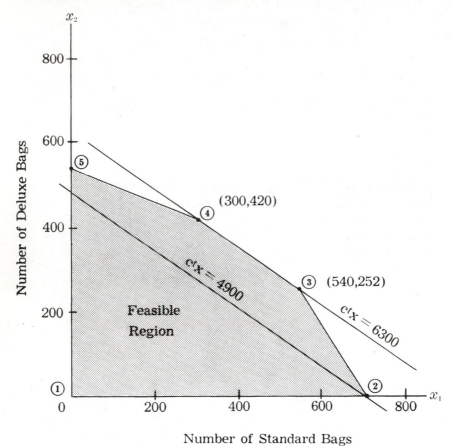

Figure 3–G.   Par, Inc. Problem with a Modified Objective
Function (Alternate Optima)

The graph of the objective function for this problem is parallel
to the cutting and dyeing constraint. As we move the profit
line away from the origin in an effort to increase the value of
the objective function, we see that the objective function is
maximized when it coincides with the cutting and dyeing constraint. The optimal objective function value for this problem
is 6300, and there are an infinite number of feasible points
yielding this value. Any point on the line segment connecting

the points (300, 420) and (540, 252) is optimal. In addition, we see that both of the end-points are optimal.

$$7(300) + 10(420) = 2100 + 4200 = 6300$$
$$7(540) + 10(252) = 3780 + 2520 = 6300$$

This condition of a linear program having alternate optima is an ideal situation for the practicing manager attempting to implement the linear programming solution. It means that several (actually an infinite number) combinations of the variables are optimal and the manager can pick the one which is most expedient.

When using the Simplex method of solution one will probably not recognize that a linear program has alternate optima until the final Simplex tableau. Then, if the program has alternate optima, $(c_j - z_j)$ will equal zero for one or more of the variables not in solution. For example, let us look at the final Simplex tableau for the problem portrayed graphically in Figure 3–G.

BASIS	c_j	x_1 7	x_2 10	s_1 0	s_2 0	s_3 0	s_4 0	
x_1	7	1	0	10/3	0	0	$-40/3$	300
s_2	0	0	0	$-10/18$	1	0	$-20/18$	100
s_3	0	0	0	$-22/9$	0	1	64/9	128
x_2	10	0	1	$-4/3$	0	0	28/3	420
z_j		7	10	10	0	0	0	6300
$c_j - z_j$		0	0	-10	0	0	0	

All values in the net evaluation row are less than or equal to zero, indicating that we have reached the optimal solution. This solution yields $x_1 = 300$, $x_2 = 420$, $s_2 = 100$ and $s_3 = 128$. Notice however, that the entry in the net evaluation row for s_4, $(c_6 - z_6)$, is equal to zero. This indicates that our linear program has alternate optima. Since $(c_j - z_j)$ for s_4 is equal to zero, we could introduce s_4 into solution without changing the value of the optimal solution. The tableau after introducing s_4 is presented below.

BASIS	c_j	x_1 7	x_2 10	s_1 0	s_2 0	s_3 0	s_4 0	
x_1	7	1	0	$-5/4$	0	120/64	0	540
s_2	0	0	0	$-30/32$	1	10/64	0	120
s_4	0	0	0	$-22/64$	0	9/64	1	18
x_2	10	0	1	15/8	0	$-84/64$	0	252
z_j		7	10	10	0	0	0	6300
$c_j - z_j$		0	0	-10	0	0	0	

After introducing s_4 we have a different solution: $x_1 = 540$, $x_2 = 252$, $s_2 = 120$, and $s_4 = 18$. However, this solution is still optimal $(c_j - z_j \leq 0$ for all j). Another way to confirm that this solution is still optimal is to note that the value of the objective function has remained at 6300.

In summary, then, we have seen that we can recognize that a linear program has alternate optima by observing graphically that the objective function is parallel to one of the binding constraints. When using the Simplex method we can recognize alternate optima if $(c_j - z_j)$ equals zero for one of the variables not in solution.

3.11   Degeneracy

A linear program is said to be degenerate if one or more of the variables in the basic solution has a value of zero. Degeneracy does not cause any particular difficulties for the graphical solution procedure; however, degeneracy can cause some difficulties when the Simplex method is used to solve a linear program.

To see how a degenerate linear program may come about consider the following modification of the Par, Inc. problem.

$$\max 10x_1 + 9x_2$$

s.t.

$$\frac{7}{10}x_1 + 1x_2 \le 630$$

$$\frac{1}{2}x_1 + \frac{5}{6}x_2 \le 480 \quad \text{(sewing capacity reduced to 480)}$$

$$1x_1 + \frac{2}{3}x_2 \le 708$$

$$\frac{1}{10}x_1 + \frac{1}{4}x_2 \le 135$$

$$x_1, x_2 \ge 0$$

Let us solve this new Par, Inc. problem using the Simplex method. The tableau after the first iteration is presented below.

BASIS	c_j	x_1 10	x_2 9	s_1 0	s_2 0	s_3 0	s_4 0	
s_1	0	0	16/30	1	0	$-7/10$	0	134.4
s_2	0	0	1/2	0	1	$-1/2$	0	126
x_1	10	1	2/3	0	0	1	0	708
s_4	0	0	22/120	0	0	$-1/10$	1	64.2
z_j		10	20/3	0	0	10	0	7080
$c_j - z_j$		0	7/3	0	0	-10	0	

The entries in the net evaluation row indicate that we should introduce variable x_2 into solution. Calculating the appropriate ratios to determine the pivot element, we get:

$$\frac{\bar{b}_1}{\bar{a}_{12}} = \frac{134.4}{16/30} = 252$$

$$\frac{\bar{b}_2}{\bar{a}_{22}} = \frac{126}{1/2} = 252$$

$$\frac{\bar{b}_3}{\bar{a}_{32}} = \frac{708}{2/3} = 1062$$

$$\frac{\bar{b}_4}{\bar{a}_{42}} = \frac{64.2}{22/120} = 350.2$$

We see that there is a tie between the first and second *rows*. This is an indication that we will have a degenerate linear program at the next iteration. To see why, let us arbitrarily select one of the tied rows and perform the necessary row operations to remove the corresponding variable from the current basis. Let us select row 1, and thus remove variable s_1 from the current basis. The Simplex tableau after this iteration is as follows:

BASIS	c_j	x_1 10	x_2 9	s_1 0	s_2 0	s_3 0	s_4 0	
x_2	9	0	1	30/16	0	$-210/160$	0	252
s_2	0	0	0	$-15/16$	1	25/160	0	0
x_1	10	1	0	$-20/16$	0	300/160	0	540
s_4	0	0	0	$-11/32$	0	45/320	1	18
	z_j	10	9	70/16	0	111/16	0	7668
	$c_j - z_j$	0	0	$-70/16$	0	$-111/16$	0	

Do you see anything unusual about this tableau? When we performed our iteration and introduced 252 units of x_2 into the basis, we not only drove s_1 out of solution setting s_1 equal to zero, but we also drove s_2 to zero. Hence, we have a solution where one of the basic variables is equal to zero. Whenever we have a tie in the $\dfrac{\overline{b}_i}{\overline{a}_{ij}}$ ratios, there will always be a basic variable equal to zero in the next tableau. Since we are at the optimal solution in this case, we do not care that s_2 is in solution at a zero value. However, if this condition were to occur at some iteration prior to reaching the optimal solution, it is theoretically possible for the Simplex algorithm to cycle. That is, the algorithm could possibly alternate between the same set of nonoptimal points at each iteration and hence never reach the optimal solution. Cycling has not proven to be a significant difficulty in practice. Therefore, we do not recommend introducing any special machinery into the Simplex algorithm to eliminate the possibility of degeneracy occurring. If while performing the iterations of the Simplex algorithm a tie occurs for the minimum $\dfrac{\bar{b}_i}{\bar{a}_{ij}}$ ratio, then we recommend simply selecting the upper row as the pivot row.

3.12 Summary

The purpose of this chapter has been to introduce the reader to extensions of the maximization type of linear program with less-than-or-equal-to constraints, the minimization linear programming problem, and some of the more important solution possibilities that can arise when solving any linear program.

We have shown in this chapter how to convert greater-than-or-equal-to constraints, equality constraints, and constraints with negative right-hand side values into the form required for writing a linear program in Standard form. Doing this required the introduction of surplus variables and noting that when both sides of an inequality are multiplied by a negative number the direction of the inequality changes. We then saw that for linear programs with greater-than-or-equal-to constraints and/or equality constraints it was necessary to introduce artificial variables in order to go from Standard to Tableau form. We assigned an objective function coefficient of $(-M)$, where M is a very large number to these variables. Thus, if there was a feasible solution to the real-world linear program, these artificial variables would be driven out of solution (or in the case of degeneracy driven to zero) before the Simplex method reached its stopping criterion.

We then presented two different techniques for solving minimization problems. The first involved changing the Simplex rules for introducing a variable into solution and changing the stopping criterion. The second approach, and the one we shall use in the remainder of the text, enabled us to solve any minimization problem using the same rules as for a maximization problem. That is, we showed that minimizing $c^t x$ was equivalent to maximizing $(-c^t x)$. The only difference was that the value of the solution to max $(-c^t x)$ was the negative of the value of the solution to min $c^t x$. Thus, to solve a minimization problem using our maximization Simplex rules, we multiply each decision variable coefficient in the objective function by -1 and then apply the maximization procedure. When we get the optimal solution to this problem, we multiply the value of the optimal solution by -1 to get the value of the optimal solution for our original minimization problem.

The remaining part of this chapter was devoted to illustrating and discussing the implications of infeasibility, unboundedness, alternate optima, and degeneracy. Toward this end, we provided a graphical illustration of the first three topics and showed the reader how to recognize and interpret all these solution possibilities when using the Simplex method.

3.13   Glossary

1. *Surplus variable*—This is a variable that is subtracted from the left-hand side of a greater-than-or-equal-to constraint in order to form an equality.

2. *Artificial variable*—A variable that has no physical meaning in terms of the original linear programming problem, but serves merely to enable a basic feasible solution to be created for starting the Simplex method. Artificial variables are assigned an objective function coefficient of $-M$, where M is a very large number.

3. *Infeasibility*—Infeasibility occurs when there is no solution to the linear programming problem which satisfies all the constraints. If the final Simplex tableau consists of one or more artificial variables in solution at a positive value, we recognize that the optimal solution to the linear programming model is infeasible for our real-world problem.

4. *Unboundedness*—A solution to a linear programming problem is said to be unbounded if the value of the solution may be made infinitely large without violating any of the constraints. If at some iteration, the Simplex method tells us to introduce x_j into solution and all of the $\bar{a}_{ij}$ are less than or equal to zero in column j, then we have an unbounded solution.

5. *Alternate optima*—When a linear program has two or more optimal solutions, we say the linear program has alternate optima. In the Simplex method we cannot usually recognize alternate optima until the final tableau. Then, if

the program has alternate optima, $(c_j - z_j)$ will equal zero for one of the variables not in solution.

6. *Degeneracy*—A linear program is degenerate if one or more of the variables in the basic solution has a value of zero.

3.14 Problems

1. Consider the following linear program

min $1.5x_1 + 2x_2$

s.t.

$$2x_1 + 2x_2 \geq 8$$
$$2x_1 + 6x_2 \geq 12$$
$$x_1, x_2 \geq 0.$$

a. Identify the feasible region and solve by the graphical procedure.

b. Solve by the Simplex method.

c. Use the graph to identify the extreme point at each iteration of the Simplex method.

2. For the above problem, use the graphical procedure and Simplex method to find the optimal solution when the objective function is

min $1.5x_1 + 1x_2$.

3. Consider the following linear program.

max $1x_1 + 2x_2$

s.t.

$$1x_1 \qquad \leq 5$$
$$1x_2 \leq 4$$
$$2x_1 + 2x_2 = 12$$
$$x_1, x_2 \geq 0$$

a. Graphically show the feasible region.

b. What are the extreme points of the feasible region?

c. Find the optimal solution by the graphical procedure and the Simplex method.

d. What is the optimal solution if the objective function is

$$\min 1x_1 + 2x_2?$$

4. Set up the Tableau form for the following linear program. (Do not attempt to solve.)

$$\min 3x_1 + 2x_2 + 4x_3 - 1x_4$$

s.t.

$$1x_1 + 1x_2 \qquad\qquad \le 40$$
$$2x_1 - 3x_2 + 3x_3 - 4x_4 \ge 10$$
$$1x_1 \qquad + 2x_3 + 1x_4 = 20$$
$$x_1, x_2, x_3, x_4 \ge 0$$

5. Set up the Tableau form for the following linear program. (Do not attempt to solve.)

$$\min 2x_1 + 5x_2 + 3x_3$$

s.t.

$$4x_1 + 2x_2 - 1x_3 = 15$$
$$3x_1 \qquad + 2x_3 \ge 10$$
$$+ 1x_2 - 1x_3 \le -5$$
$$- 2x_1 + 2x_2 \qquad = -8$$
$$x_1, x_2, x_3 \ge 0$$

6. Solve the following linear program.

$$\min 3x_1 + 4x_2 + 8x_3$$

s.t.

$$2x_1 + 1x_2 \qquad \ge 6$$
$$2x_2 + 4x_3 \ge 8$$
$$x_1, x_2, x_3 \ge 0$$

7. Solve the following linear program.

$$\min 4x_1 + 2x_2 + 3x_3$$

s.t.

$$1x_1 + 3x_2 \qquad \ge 15$$
$$1x_1 \qquad + 2x_3 \ge 10$$
$$2x_1 + 1x_2 \qquad \ge 20$$
$$x_1, x_2, x_3 \ge 0$$

8. Doc's Dog Kennels, Inc. provides overnight lodging for a variety of pets. A particular feature at Doc's is the quality of care the pets receive, including excellent food. The kennel's dog food is made by mixing two brand name dog food products to obtain what Doc's calls the "well-balanced dog diet". The data for the two dog foods are shown below:

Dog Food	Cost/oz.	% Protein	% Fat
Bark Bits	$.03	30	10
Canine Chow	$.025	20	20

If Doc wants to be sure that his dogs receive at least 5 ounces of protein and 2 ounces of fat per day, what is the minimum cost mix of the two dog food products?

a. Solve by the graphical procedure.

b. Solve by the Simplex method.

c. What is the verbal interpretation of the surplus variables and their values?

9. Jack Kammer has been trying to figure out the correct amount of fertilizer that should be applied to his lawn. After getting his soil analyzed at the local argricultural agency, he has been advised to put at least 60 pounds of nitrogen, 24 pounds of phosphorous compounds, and 40 pounds of potassium compounds on the lawn this season. One-third of the mixture is to be applied in May, one-third in July, and one-third in late September. After checking the local discount stores, Jack finds that one store is currently having a sale on packaged fertilizer. One type on sale is the 20–5–20 mixture consisting of 20% nitrogen, 5% phosphorous compounds, and 20% potassium compounds, and selling at $4 for a 20 pound bag. The other type on sale is a 10–10–5 mixture selling for $5 for a 40 pound bag. Jack would like to know how many bags of each type he should purchase so he can mix the ingredients together to form a mixture that will meet the minimum agricultural agency requirements. Like all homeowners that are plagued with large lawns, Jack would like to spend as little as he needs to in order to keep his lawn healthy. What should Jack do?

What is your interpretation of the surplus variables and their values?

10. Ajax Fuels, Inc. is developing a new additive for airplane fuels. The additive is a mixture of three liquid ingredients (A, B, and C). For proper performance, the total amount of additive must be at least 10 ounces for each gallon of fuel. However, because of safety reasons the amount of additive should not exceed 15 ounces in each gallon of fuel. The mix of ingredients in the additive is also critical. At least ¼ of an ounce of ingredient A must be used for every ounce of ingredient B and at least 1 ounce of ingredient C must be used for every ounce of ingredient A. If the cost of the ingredients (A, B, and C) are $.10, $.03, and $.09 per ounce, respectively, find the minimum-cost mixture of A, B, and C for the additive and the amount of the additive that should be used for every gallon of airplane fuel.

Note: In problems 11 to 19, we provide example linear programs that result in one or more of the following solution situations:

    1. optimal solution
    2. infeasible solution
    3. unbounded solution
    4. alternate optimal solution
    5. degenerate solution.

For each linear program, define the solution situations that exist and indicate how you identified each situation in the Simplex tableau. For the problems with alternate optimal solutions, calculate at least two optimal solutions.

11.   min $1x_1 + 1x_2 + 1x_3$
    s.t.

$$1x_1 + 5x_2 + 2x_3 \geq 250$$
$$1x_2 + 2x_3 \geq 50$$
$$x_1, x_2, x_3 \geq 0$$

12.   max $2x_1 + 4x_2$
      s.t.

$$1x_1 + 1x_2 \leq 5$$
$$-1x_1 + 1x_2 \geq 8$$
$$x_1,\ x_2 \geq 0$$

Also solve problem 12 using the graphical procedure.

13.   max $1x_1 + 1x_2$
      s.t.

$$4x_1 + 3x_2 \geq \ \ 12$$
$$2x_1 + 3x_2 \geq -\ 6$$
$$1x_2 \geq \ \ 2$$
$$x_1,\ x_2 \geq 0$$

Also solve problem 13 using the graphical procedure.

14.   min $3x_1 + 3x_2$
      s.t.

$$4x_1 + 1x_2 \geq 20$$
$$1x_1 \qquad\ \geq \ \ 2$$
$$2x_1 + 2x_2 \geq 16$$
$$x_1,\ x_2 \geq 0$$

Also solve problem 14 using the graphical procedure.

15.   min $-\ 4x_1 + 5x_2 + 5x_3$
      s.t.

$$-\ 1x_2 + 1x_3 \geq 2$$
$$-\ 1x_1 + 1x_2 + 1x_3 \geq 1$$
$$x_3 \leq -\ 1$$
$$x_1,\ x_2,\ x_3 \geq 0$$

16.   min $2x_1 + 1x_2$
      s.t.

$$1x_1 + 1x_2 = 600$$
$$1x_1 + 1x_2 \leq 700$$
$$1x_1 + \tfrac{3}{2}x_2 \leq 900$$
$$1x_1 \qquad\ \geq 200$$
$$x_1,\ x_2 \geq 0$$

17. max $2x_1 + 4x_2$

    s.t.

$$1x_1 + \tfrac{1}{2}x_2 \leq 10$$
$$1x_1 + 1x_2 = 12$$
$$1x_1 + \tfrac{3}{2}x_2 \leq 18$$
$$x_1, x_2 \geq 0$$

18. max $1x_1 + 2x_2 + 1x_3$

    s.t.

$$3x_1 + 4x_2 \qquad \leq 12$$
$$2x_1 + 3x_2 - 1x_3 \geq 6$$
$$x_1, x_2, x_3 \geq 0$$

19. max $2x_1 + 1x_2 + 1x_3$

    s.t.

$$2x_1 + 1x_2 + 1x_3 \geq 2$$
$$1x_1 + 2x_2 \qquad \leq 10$$
$$2x_1 + 4x_2 + 1x_3 \leq 8$$
$$x_1, x_2, x_3 \geq 0$$

20. In section 3.1, we defined a modified Par, Inc. problem where the minimum production levels for each golf bag were given by

$$x_1 \geq 100$$
$$x_2 \geq 100.$$

While it was perfectly acceptable to use these constraints in the Simplex procedure, consider the following variation. Since we know x_1 and x_2 will both be at least 100, let us define new decision variables x_1' and x_2' where

$x_1' =$ production of Standard bags above the 100-unit minimum

$x_2' =$ production of Deluxe bags above the 100-unit minimum.

Thus, if we know the values of x_1' and x_2', we can find our total production x_1 and x_2 by

$$x_1 = 100 + x_1',$$
$$x_2 = 100 + x_2'.$$

Return to the Par, Inc. problem in section 3.1 and substitute the expression $x_1 = 100 + x_1'$ and $x_2 = 100 + x_2'$ into the linear program. State the linear program in terms of the x_1 and x_2 decision variables.

What is the primary advantage of this procedure? Solve the linear program for the optimal value of the x_1' and x_2' variables. Do we obtain the same optimal production plan?

21. In addition to their line of bicycles, Hot Wheels, Inc. manufactures three types of kiddie tricycles. They produce a model known as the Fat Wheel, a model called the Toad, and their ever popular model, the Ridge Runner. Hot Wheels manufactures these tricycle models on special order or whenever they have any slack time available during their bicycle production. Hot Wheels currently has some slack time available, and would like to determine the optimal number of kiddie tricycles to produce in order to maximize the total number of tricycles produced, with the requirement that they produce at least twice as many Ridge Runners as they do of Fat Wheels and Toads. (Historically, orders have normally been 2–1 in favor of the Ridge Runner Model.) Each unit of the Fat Wheel and Toad requires 10 minutes of manufacturing time, whereas each unit of the Ridge Runner requires 4 minutes. In addition, Fat Wheels require 8 minutes of assembly time, Toads require 6 minutes, and Ridge Runners require 4 minutes. There are 40 hours of manufacturing time available and 20 hours of assembly time available. The warehouse has capacity to store a maximum of 150 tricycles. What should Hot Wheels do?

Consider the possibility of alternate optimal solutions. What flexibility does this provide for Hot Wheels?

22. Supersport Football's, Inc. has just received a rush order for 1000 of their All-Pro model footballs. Because Supersport is the sole manufacturer of NFL footballs, the current production run must result in at least 1000 All-Pro models being produced. Supersport also manufactures a College Model and a High-School Model. All three footballs require operations in the following departments: cutting and dyeing, sewing, and inspection and packaging.

The production times and maximum production availabilities are shown below:

| | Production Operations | | |
Model	Cutting & Dyeing	Sewing	Inspection & Packaging
All-Pro	12	15	3
College	10	15	4
High-School	8	12	2
Time Available	300 hrs.	200 hrs.	100 hrs.

a. If Supersport realizes a profit of $3.00 for each All-Pro model, $5.00 for each College model, and $4.00 for each High-School model, how many footballs of each type should be produced?

b. If Supersport can increase Sewing time to 300 hours and Inspection and Packaging time to 150 hours by using overtime, what is your recommendation?

23. Captain John's Yachts, Inc., located in Fort Lauderdale, Florida, rents three types of ocean-going boats: sailboats, cabin cruisers, and Captain John's favorite—the luxury yachts. Captain John advertises his boats with his famous "you rent—we pilot" slogan, which means the company supplies the captain and crew for each rented boat. Each rented boat, of course, has one captain, but the crew sizes (i.e., deck hands, galley, etc.) differ. The crew requirements, in addition to a captain, are 1 for sailboats, 2 for cabin cruisers, and 3 for yachts. Currently, Captain John has rental requests for all of his boats: 4 sailboats, 8 cabin cruisers, and 3 luxury yachts. However, he only has 10 employees who qualify as captains and an additional 18 employees who qualify for the crew positions. If Captain John's daily profit is $50 for sailboats, $70 for cruisers, and $100 for luxury yachts, how many boats of each type should he rent?

24. The cook at Happy Harry's Lakeside Kiddie Resort has a problem. In addition to knowing how to cook just three different dishes, he has been told by Happy Harry to use

as much as possible of the ingredients on hand in order to make up meals having the highest nutritional value possible (this is so Harry can advertise that his resort is not only a fun place to get rid of the kids, but also a healthy place). The cook currently has 40 pounds of ingredient A available, 30 pounds of ingredient B, and 60 pounds of ingredient C. Each unit of recipe 1 calls for 1 pound of A, ½ pound of B, and 1 pound of C. Each unit of recipe 2 requires 2 pounds of B and 1 pound of C. Each unit of recipe 3 requires 1 pound of both A and B, and 2 pounds of C. If one unit of recipe 1 contains 15 nutritional units, one unit of recipe 2 contains 30 nutritional units, and one unit of recipe 3 contains 25 nutritional units, how many units of each recipe should the cook make in order to maximize the nutritional value of the meals made. Set up this problem and solve using the Simplex method.

25. The We-Survey-Anything Marketing Research Company has just been hired by Ace Industries to investigate consumer reaction to Ace's newly introduced product. The contract calls for a door-to-door survey with the following stipulations:

1. At least 200 households with no children be contacted, either during the day or evening;

2. At least 400 households with children be contacted, either during the day or evening;

3. The total number of households contacted during the evening be at least as great as the number contacted during the day;

4. A sample of at least 1,000 families be contacted during the study.

Based upon previous interviews management has developed the following interview costs:

		Interview	
		Day	Evening
Household	Children	$10	$12
	No Children	$ 8	$10

What household-time of day plan should the company use in order to minimize interview costs while satisfying contract requirements?

26. The Our-Paint-Dries-Quickest (OPDQ) Paint Company produces two interior enamels: Quick-Dry and Super-Speedie. Both enamels are manufactured from pre-mix silicate base and linseed oil solutions, which OPDQ purchases from a number of different suppliers. Currently, only two types of pre-mix solutions are available. Type A contains 60% silicates and 40% linseed oil, whereas Type B contains 30% silicates and 70% linseed oil. Type A costs $0.50 per gallon and Type B costs $0.75 per gallon. If each gallon of Quick-Dry requires at least 25% silicates and 50% linseed oil, and each gallon of Super-Speedie requires at least 20% silicates but at most 50% linseed oil, how many gallons of each pre-mix should OPDQ purchase in order to produce exactly 100 gallons of Quick-Dry and 100 gallons of Super-Speedie?

4

Duality and Sensitivity Analysis

In the last chapter we studied minimization problems and some of the special cases that might occur when we attempt to solve linear programming problems. We are now ready to extend our knowledge and understanding of linear programming to include the topics of duality and sensitivity analysis.

In studying duality we will learn about two related linear programs: the Primal and the Dual. We shall see that every linear programming problem can be written as a Primal problem and that for every Primal problem there exists a corresponding linear programming problem called the Dual problem. It turns out that the solution to either of these problems also provides the solution to the other. Thus one important reason for studying duality is that the solution to every linear programming problem can be found by solving either a Primal or a Dual problem. If the Primal and Dual problems differ in terms of computational difficulty, we can simply choose the easier one to solve.

Duality also plays a very important role in the area of sensitivity analysis. We shall discuss this phase of linear programming later in the chapter. For now, let us begin our study by defining exactly what we mean by Primal and Dual linear programming problems.

4.1   The Primal Problem

The *Primal problem* is a linear programming problem having the following properties:

1. the objective function involves maximization

2. the constraints are *all* less-than-or-equal-to constraints. Thus, in matrix notation, the Primal problem is written as follows: [1]

$$\max c^t x$$
$$\text{s.t.}$$
$$Ax \leq b$$
$$x \geq 0.$$

where

c = column vector containing the coefficients in the objective function,

x = column vector containing the decision variables,

A = matrix containing the coefficients in the constraints,

b = column vector containing the right-hand side values of the constraints.

Before defining the Dual problem, let us show how any linear programming problem can be written in the form of the Primal problem. Since the Primal problem has a maximization objective function and only less-than-or-equal-to constraints, we will need to know how to convert minimization objective functions, greater-than-or-equal-to constraints, and equality constraints into equivalent expressions appropriate for Primal form.

We learned, in the last chapter, an easy way to transform a minimization problem into an equivalent maximization problem: simply multiply the objective function by (-1). The same values

---

[1] Some authors in developing duality relationships consider a variety of forms for the Primal problem and then derive a different Dual problem for each. We believe that it is best for a text at this level to consider only one general form of the Primal problem and then adopt standard procedures for converting the Primal problem to the Dual problem.

of the decision variables which solve the maximization problem will also solve the minimization problem. The only difference is that the value of the optimal solution to the transformed maximization problem is the negative of the value of the optimal solution to the original minimization problem. Hence, we see that we can easily convert any minimization problem into an equivalent maximization linear program.

Greater-than-or-equal-to constraints can also be easily converted into the required Primal form. Consider for example the constraint:

$$4x_1 + 3x_2 \geq 7.$$

We can change this to a less-than-or-equal-to constraint by simply multiplying by (-1). Doing this we get:

$$-4x_1 - 3x_2 \leq -7.$$

Wait a minute! We learned earlier that all of the values on the right-hand side of our constraints must be nonnegative if we are going to apply the Simplex method of solution. Well, we should not be concerned with that difficulty now. We are merely interested in stating the problem as a Primal problem so that we may find its Dual. Later, if we attempt to solve the Primal or the Dual problem using the Simplex method, we will indeed have to ensure that all the right-hand side values are nonnegative.

The procedure for converting equality constraints into the less-than-or-equal-to constraints required for the Primal problem is also quite simple. Let us illustrate the procedure using the following equality constraint:

$$6x_1 - 3x_2 + 4x_3 = 27.$$

To convert this equality to inequality form, all we do is replace the single equality constraint with *two* inequality constraints: one a less-than-or-equal-to constraint, and the other a greater-than-or-equal-to constraint. Doing this for the above equation we get:

$$6x_1 - 3x_2 + 4x_3 \geq 27$$
$$6x_1 - 3x_2 + 4x_3 \leq 27.$$

Obviously, the only way both of the above constraints can be satisfied is if the original equality constraint is satisfied. And, conversely, if the original equality constraint is satisfied, both of the above inequality constraints must be satisfied. Of course the Primal problem requires that we have all less-than-or-equal-to constraints; hence, we must multiply the first inequality above by (-1). Our original equality constraint can now be replaced by the two less-than-or-equal-to constraints:

$$-6x_1 + 3x_2 - 4x_3 \leq -27$$
$$6x_1 - 3x_2 + 4x_3 \leq \quad 27.$$

As an illustration of what we have been discussing, let us write the following linear program in its Primal form.

$$\min 2x_1 + 3x_2$$
s.t.
$$6x_1 - 1x_2 = 10$$
$$4x_1 - 2x_2 \geq \quad 3$$
$$x_1, x_2 \geq 0$$

As a first step we convert the equality constraints into two inequalities to obtain the following problem:

$$\min 2x_1 + 3x_2$$
s.t.
$$6x_1 - 1x_2 \geq 10$$
$$6x_1 - 1x_2 \leq 10$$
$$4x_1 - 2x_2 \geq \quad 3$$
$$x_1, x_2 \geq 0.$$

We next convert the $\geq$ constraints to $\leq$ constraints by multiplying through by (-1). In addition, we multiply the objective function by (-1) to obtain a maximization problem. This gives us the following Primal problem:

$$\max -2x_1 - 3x_2$$
s.t.
$$-6x_1 + 1x_2 \leq -10$$
$$6x_1 - 1x_2 \leq \quad 10$$
$$-4x_1 + 2x_2 \leq -\ 3$$
$$x_1, x_2 \geq 0.$$

By using the above procedures, any linear programming problem can be written as an equivalent Primal problem. The above Primal problem is equivalent to the original linear programming problem in that the optimal solution to the Primal form of the problem and the optimal solution to the original form of the problem will be the same.

4.2   The Dual Problem

We have stated that every Primal problem has a corresponding *Dual problem*. In matrix notation, the Dual problem is written

$$\min b^t u$$

s.t.

$$A^t u \geq c$$

$$u \geq 0.$$

where

$b^t =$ transpose of the right-hand side (b vector) of the Primal problem,

$A^t =$ transpose of the coefficient matrix A of the Primal problem,

$c =$ the objective function coefficients (c vector) of the Primal problem,

$u =$ the vector of dual variables.

Note that the Dual problem as presented above is also a linear programming problem; however, the Dual problem is a minimization problem subject to $\geq$ constraints, whereas the Primal problem is a maximization problem subject to $\leq$ constraints. We also note that the coefficients of the objective function in the Dual (b^t) are the values that appear on the right-hand side of the Primal constraints (b), and the coefficients of the Primal objective function (c^t), are the values that appear on the right-hand side of the dual constraints (c). Furthermore, the matrix of coefficients for the Dual (A^t) is just the transpose of the matrix of coefficients for the Primal (A). The other major difference in the two problems is that the original variables x from the Primal problem have been replaced by the dual variables u for the Dual problem. The interpretation of these new dual variables will be discussed later in this chapter. These relationships be-tween the Primal and the Dual are part of the definition of the two problems and hence will always hold.

Let us now return to our example problem. In matrix repre-sentation, the Primal problem was written as

$$\max \ c^t x$$

s.t.

$$Ax \leq b$$

$$x \geq 0$$

where

$$c^t = [-2 - 3],$$

$$x = \begin{bmatrix} x_1 \\ x_2 \end{bmatrix}, \qquad b = \begin{bmatrix} -10 \\ 10 \\ -3 \end{bmatrix}$$

and

$$A = \begin{bmatrix} -6 & 1 \\ 6 & -1 \\ -4 & 2 \end{bmatrix}$$

In defining the Dual problem, we must use

$$c = \begin{bmatrix} -2 \\ -3 \end{bmatrix} \quad b^t = [-10 \quad 10 - 3], \text{ and } A^t = \begin{bmatrix} -6 & 6 & -4 \\ 1 & -1 & 2 \end{bmatrix}$$

Since the objective function will be $b^t u$, where b^t has three ele-ments, and since the constraints will be $A^t u \geq c$, where A^t has three columns, you can see that in order to conform for matrix multiplication, u must have three elements. Thus, the dual variables are

$$u = \begin{bmatrix} u_1 \\ u_2 \\ u_3 \end{bmatrix}$$

Carrying out the vector and matrix multiplication rules for the Dual

$$\min \ b^t u$$
$$\text{s.t.}$$
$$A^t u \geq c$$
$$u \geq 0.$$

we have

$$\min \ -10u_1 + 10u_2 - 3u_3$$
$$\text{s.t.}$$
$$-6u_1 + \ 6u_2 - 4u_3 \geq -2$$
$$1u_1 - \ 1u_2 + 2u_3 \geq -3$$
$$u_1, \ u_2, \ u_3 \geq 0.$$

As another illustration of how we can develop the Dual form of a linear programming problem, consider the Par, Inc. problem as shown below:

$$\max \ 10x_1 + 9x_2$$
$$\text{s.t.}$$
$$\tfrac{7}{10}x_1 + 1x_2 \leq 630$$
$$\tfrac{1}{2}x_1 + \tfrac{5}{6}x_2 \leq 600$$
$$1x_1 + \tfrac{2}{3}x_2 \leq 708$$
$$\tfrac{1}{10}x_1 + \tfrac{1}{4}x_2 \leq 135$$
$$x_1, \ x_2 \geq 0.$$

Since this is a maximization problem with all less-than-or-equal-to constraints, the problem is already in its Primal form with

$$c^t = [10 \quad 9]$$

$$b = \begin{bmatrix} 630 \\ 600 \\ 708 \\ 135 \end{bmatrix} \quad \text{and} \quad A = \begin{bmatrix} \tfrac{7}{10} & 1 \\ \tfrac{1}{2} & \tfrac{5}{6} \\ 1 & \tfrac{2}{3} \\ \tfrac{1}{10} & \tfrac{1}{4} \end{bmatrix}$$

For the Dual problem, we define the following:

$$c = \begin{bmatrix} 10 \\ 9 \end{bmatrix} \quad , \quad b^t = [630 \quad 600 \quad 708 \quad 135],$$

$$A^t = \begin{bmatrix} \frac{7}{10} & \frac{1}{2} & 1 & \frac{1}{10} \\ 1 & \frac{5}{6} & \frac{2}{3} & \frac{1}{4} \end{bmatrix}$$

and

$$u = \begin{bmatrix} u_1 \\ u_2 \\ u_3 \\ u_4 \end{bmatrix}$$

Thus, we have the following Dual for the Par, Inc. problem.

$$\min 630u_1 + 600u_2 + 708u_3 + 135u_4$$

s.t.

$$\frac{7}{10}u_1 + \frac{1}{2}u_2 + 1u_3 + \frac{1}{10}u_4 \geq 10$$
$$1u_1 + \frac{5}{6}u_2 + \frac{2}{3}u_3 + \frac{1}{4}u_4 \geq 9$$
$$u_1, u_2, u_3, u_4 \geq 0.$$

Let us now discuss the relationship between the Primal and Dual in terms of the number of variables and the number of constraints. Suppose the Primal problem consists of n variables and m constraint inequalities. Since the coefficients of the objective function of the Dual problem are the m right-hand side values of the Primal problem, the Dual problem must consist of m variables. In addition, since the dual right-hand sides are the coefficients of the objective function in the Primal problem, the Dual problem must have n constraints. For example, note that the two-variable, four-constraint Primal form of the Par, Inc. problem resulted in a four-variable, two-constraint Dual problem.

Another way of looking at the relationship between the number of variables and number of constraints is to think of each dual variable as being associated with one constraint in the Primal problem. That is, the first dual variable is associated with the first primal constraint, the second dual variable is associated with the second primal constraint, and so on. Thus, in the Par, Inc. problem, we needed four dual variables for the four primal constraints. The first dual variable, u_1, is associated with the cutting and dyeing constraint, $\frac{7}{10}x_1 + 1x_2 \leq 630$; the second dual variable, u_2, is associated with the sewing constraint, $\frac{1}{2}x_1 + \frac{5}{6}x_2 \leq 600$; etc.

At this point, you should be able to set up the Dual problem corresponding to any Primal linear programming problem. This can be accomplished by the following two-step procedure:

1. transform the original linear programming problem into its equivalent Primal problem;

2. apply the definition of the Dual problem given at the beginning of this section.

4.3 Properties of the Dual and Its Relationship to the Primal

We stated in the introduction to this chapter that solving either the Primal or Dual problem will provide the solution to the other. Let us solve the Dual of the Par, Inc. problem and see what observations we can make about the relationship between the Primal and Dual. Recall that the Dual for the Par, Inc. problem was

$$\min 630u_1 + 600u_2 + 708u_3 + 135u_4$$

s.t.

$$\frac{7}{10}u_1 + \frac{1}{2}u_2 + 1u_3 + \frac{1}{10}u_4 \geq 10$$

$$1u_1 + \frac{5}{6}u_2 + \frac{2}{3}u_3 + \frac{1}{4}u_4 \geq 9$$

$$u_1, u_2, u_3, u_4 \geq 0.$$

Multiplying the objective function by (-1) to convert to a maximization problem, subtracting surplus variables s_1 and s_2

from both equations, and adding artificial variables a_1 and a_2, we can write the first Simplex tableau as follows:

BASIS	c_j	u_1 -630	u_2 -600	u_3 -708	u_4 -135	s_1 0	s_2 0	a_1 $-M$	a_2 $-M$	
a_1	$-M$	$7/10$	$1/2$	1	$1/10$	-1	0	1	0	10
a_2	$-M$	$\boxed{1}$	$5/6$	$2/3$	$1/4$	0	-1	0	1	9
z_j		$-\dfrac{17M}{10}$	$-\dfrac{4M}{3}$	$-\dfrac{5M}{3}$	$-\dfrac{42M}{120}$	M	M	$-M$	$-M$	$-19M$
c_j-z_j		$-630+\dfrac{17M}{10}$	$-600+\dfrac{4M}{3}$	$-708+\dfrac{5M}{3}$	$-135+\dfrac{42M}{120}$	$-M$	$-M$	0	0	

By introducing variable u_1 at the first iteration, we drive out a_2. Performing the usual Simplex operations, the second tableau becomes:

BASIS	c_j	u_1 -630	u_2 -600	u_3 -708	u_4 -135	s_1 0	s_2 0	a_1 $-M$	a_2 $-M$	
a_1	$-M$	0	$-1/12$	$8/15$	$-9/120$	-1	$\boxed{7/10}$	1	$-7/10$	$3\tfrac{7}{10}$
u_1	-630	1	$5/6$	$2/3$	$1/4$	0	-1	0	1	9
z_j		-630	$-525+\dfrac{M}{12}$	$-420-\dfrac{8M}{15}$	$-157\tfrac{1}{2}+\dfrac{9M}{120}$	M	$630-\dfrac{7M}{10}$	$-M$	$\dfrac{7M}{10}-630$	$\dfrac{-37M}{10}-5670$
c_j-z_j		0	$-75-\dfrac{M}{12}$	$-288+\dfrac{8M}{15}$	$22\tfrac{1}{2}-\dfrac{9M}{120}$	$-M$	$\dfrac{7M}{10}-630$	0	$630-\dfrac{17M}{10}$	

At the second iteration, we introduce s_2, and a_1 is driven out of solution. After performing the Simplex operations, our third tableau becomes:

BASIS	c_j	u_1 -630	u_2 -600	u_3 -708	u_4 -135	s_1 0	s_2 0	a_1 $-M$	a_2 $-M$	
s_2	0	0	$-10/84$	$\boxed{16/21}$	$-9/84$	$-10/7$	1	$10/7$	-1	$37/7$
u_1	-630	1	$60/84$	$30/21$	$12/84$	$-10/7$	0	$10/7$	0	$100/7$
z_j		-630	-450	-900	-90	900	0	-900	0	-9000
c_j-z_j		0	-150	192	-45	-900	0	$900-M$	$-M$	

Notice that all of the artificial variables have now been driven out and we have a basic feasible solution to our Dual (i. e., $s_2 = {}^{35}/_7$ and $u_1 = {}^{100}/_7$). At the third iteration, we introduce u_3, and s_2 is driven out of solution. After performing the Simplex operations, our fourth tableau becomes:

BASIS	c_j	u_1 -630	u_2 -600	u_3 -708	u_4 -135	s_1 0	s_2 0	a_1 $-M$	a_2 $-M$	
u_3	-708	0	$-5/32$	1	$-9/64$	$-15/18$	$21/16$	$15/8$	$-21/16$	$111/16$
u_1	-630	1	$15/16$	0	$11/32$	$5/4$	$-15/8$	$-5/4$	$15/8$	$35/8$
	z_j	-630	-480	-708	-117	540	252	-540	-252	-7668
	c_j-z_j	0	-120	0	-18	-540	-252	$540-M$	$252-M$	

Since all the coefficients in the net evaluation row are ≤ 0, this iteration has provided us with the optimal solution. The optimal solution for the Dual problem is

$$u_1 = {}^{35}/_8, \quad u_2 = 0, \quad u_3 = {}^{111}/_{16}, \quad u_4 = 0$$
$$s_1 = 0, \quad s_2 = 0, \quad a_1 = 0, \text{ and } a_2 = 0.$$

Since we have been maximizing the negative of our dual objective function, the optimal value of the objective function for the Dual problem must be $-(-7668)$ or 7668.

Shown below is the final Simplex tableau for the Primal form of the Par, Inc. problem.

BASIS	c_j	x_1 10	x_2 9	s_1 0	s_2 0	s_3 0	s_4 0	
x_2	9	0	1	$30/16$	0	$-210/160$	0	252
s_2	0	0	0	$-15/16$	1	$25/160$	0	120
x_1	10	1	0	$-20/16$	0	$300/160$	0	540
s_4	0	0	0	$-11/32$	0	$45/320$	1	18
	z_j	10	9	$70/16$	0	$111/16$	0	7668
	c_j-z_j	0	0	$-70/16$	0	$-111/16$	0	

The optimal solution for the Primal problem is

$$x_1 = 540, \; x_2 = 252, \; s_1 = 0, \; s_2 = 120, \; s_3 = 0, \text{ and } s_4 = 18.$$

The optimal value of the objective function is 7668.

What observations can you make about the relationship between the solutions to the Primal and Dual of the Par, Inc. problem? First, we note that the optimal value of the objective function is the same (7668) for both. This is true for all Primal and
Dual linear programming problems and is stated as Property 1
below:

Property 1. If the Dual problem has an optimal solution, the
Primal problem has an optimal solution and vice versa. Furthermore, the values of the optimal solutions to the Dual and Primal
problems are equal.

This property tells us that if we had solved only the Dual problem, we would have known that Par, Inc. could make a maximum
of $7668 from the production of Standard and Deluxe golf bags.

Now can we use the solution to the Dual problem to tell us how
many Standard bags (x_1) and how many Deluxe bags (x_2) Par,
Inc. should produce? That is, can we use the results from the
Dual to determine the optimal values for the Primal problem
decision variables x_1 and x_2? Look closely at the net evaluation
row in the final Simplex tableau for the Dual problem. Can you
identify the $x_1 = 540$ and $x_2 = 252$ optimal solution values for the
Primal problem? Your answer should be "yes" since the values
of the primal variables are simply the negative of the $c_j - z_j$
values for the surplus variables in the Dual problem. That is,

Surplus Variable

$$s_1 \longrightarrow x_1 = -(c_5 - z_5) = 540,$$
$$s_2 \longrightarrow x_2 = -(c_6 - z_6) = 252.$$

Note also that the negative of the net evaluation row values
$[-(c_j - z_j)]$ for the slack variables in the final Simplex tableau

for the Primal problem provides the values of the dual variables. That is,

Slack Variable

$$s_1 \longrightarrow u_1 = -(c_3 - z_3) = {}^{70}\!/_{16} = {}^{35}\!/_{8}$$

$$s_2 \longrightarrow u_2 = -(c_4 - z_4) = 0$$

$$s_3 \longrightarrow u_3 = -(c_5 - z_5) = {}^{111}\!/_{16}$$

$$s_4 \longrightarrow u_4 = -(c_6 - z_6) = 0.$$

This procedure for determining the solution to the Primal problem from the net evaluation row of the final Simplex tableau of the Dual problem and vice versa is Property 2 of the relationship between the Primal and Dual linear programs.

Property 2. The optimal values of the primal variables are given by the negative of the $(c_j - z_j)$ entries for the surplus [1] variables in the Simplex tableau corresponding to the optimal Dual solution.

Similarly, the optimal values of the dual variables are given by the negative of the $(c_j - z_j)$ entries for the slack variables in the Simplex tableau corresponding to the optimal Primal solution. Thus, Property 2 tells us that if we had solved only the Dual problem, we still could have determined the optimal values of our decision variables ($x_1 = 540$, $x_2 = 252$) for the Primal problem by using the net evaluation row of the final Simplex tableau.

---

[1] Since the Dual problem has all $\geq$ constraints, we must use surplus variables to obtain Tableau form for the Dual problem. Suppose the second constraint of a Dual problem were written $3u_1 - 2u_2 \geq -5$. Technically, we subtract a surplus variable s_2 to get $3u_1 - 2u_2 - s_2 = -5$ and the optimal value of primal variable x_2 is found from the $-(c_j - z_j)$ corresponding to s_2. However, in setting up the initial Simplex tableau, we will satisfy the nonnegative right-hand side requirement by writing the above constraint as $-3u_1 + 2u_2 + s_2 = 5$. In such cases, s_2 will have the appearance of a slack variable in the Simplex tableau (i. e., a coefficient of $+1$). Whether you prefer to call s_2 a slack variable or a surplus variable for the Dual problem is up to you; the important point is that you realize x_2 is given by the $-(c_j - z_j)$ for variable s_2 in the Simplex tableau corresponding to the optimal Dual solution.

A similar argument is applicable for the column associated with the slack variable when one of the right-hand side values for the Primal problem is negative.

What other information can we obtain by solving the Dual problem? Recall that the interpretation of the slack variables in the Primal Par, Inc. problem told us which operations were restricting our optimal solution. Specifically, we saw that $s_1 = 0$ and $s_3 = 0$. This told us that there was no slack time in either the cutting and dyeing department or the finishing department. Thus, the corresponding constraints were binding.

This same information can be obtained from the Dual solution based on Property 3 which is called the principle of *complementary slackness*.

Property 3. If a primal constraint is satisfied as a strict inequality [2] in the optimal solution, then the corresponding dual variable will be equal to zero in the optimal Dual solution. Conversely, if a dual variable is strictly positive (>0) in the optimal solution to the Dual problem, then the corresponding primal constraint will be satisfied as an equality in the optimal solution to the Primal problem.

Thus, given these three duality properties, we can see that using the Simplex method to solve the Dual linear programming problem provides much of the same information that is contained in the solution to the Primal linear programming problem. Specifically, if the Primal problem has an optimal solution, the optimal solution to the Dual problem provides

1. the optimal value for the objective function of the Primal problem,

2. the optimal values of the decision variables for the Primal problem, and

3. the constraints that are satisfied as equalities and the constraints that may be satisfied as strict inequalities in the Primal problem.

These properties are very important from a practical point of view. They imply that once we have solved either the Primal or

---

[2] We say that a less-than-or-equal-to constraint is satisfied as a strict inequality whenever the value on the left-hand side of the constraint is less than the value on the right-hand side. For example, the second constraint in the Par, Inc. problem is satisfied as a strict inequality at the optimal solution since $(\frac{1}{2})(540) + (\frac{5}{6})(252) = 480$ is less than 600, the original right-hand side value.

Dual problem, we have solved the other as well. Hence, we should pick the easiest one to solve. Recall that if the Primal problem has m constraints, then the Dual problem has m variables, and vice versa. In addition, we know that any basic solution has just as many basic variables in it as there are constraints in the problem. Therefore, the more constraints there are in the problem the more basic variables there will be in solution at each iteration of the Simplex method. For most computer procedures designed for large-scale linear programming problems, the number of computations performed at each iteration and the total number of iterations are closely related to the number of basic variables. Hence, the number of basic variables may significantly influence the computation time for the solution procedure.

Thus, in general, we expect linear programs with a large number of constraints, and therefore a large number of basic variables, to require greater computational time and effort. When computational time is an important and/or expensive factor in obtaining a solution, practitioners of linear programming usually recommend solving the form of the problem (Primal or Dual) that has the fewer number of constraints.

For example, suppose the Primal linear programming problem had $400 \leq$ constraints and 80 decision variables. Counting slack variables, the Primal problem would have a total of 480 variables and 400 constraints when placed in Tableau form. The Dual problem, counting the surplus and artificial variables necessary for Tableau form, would have 560 variables, but only 80 constraints. Hence, this is a situation in which it would clearly be easier to solve the Dual problem. Other cases are not nearly so obvious, but a general rule of thumb is that the problem with the fewer constraints is the easier to solve.

To complete this section on duality, we present two additional properties relating the Primal and Dual. One of these properties states that a feasible solution to the Dual problem always provides a value for the objective function that is greater than or equal to the value of the objective function for any feasible solution to the Primal problem. That is, if we have any feasible solution to the Dual problem, and any feasible solution to the Primal problem, the value of the dual objective function would be greater than or equal to the value of the primal objective function. For example, the third tableau in the Dual solution procedure contains a dual feasible solution with an objective function value of $9000 = -(-9000)$. Clearly, this value is greater than or equal to the value of the objective function for any feasible Primal solution since the value of the maximizing solution to the Primal problem is 7668. We may now state Property 4 as follows:

Property 4. The value of the objective function for any feasible solution to the Dual problem is greater than or equal to the value of the objective function for any feasible solution to the Primal problem.

The next property is applicable to the case where there is no optimal solution to either the Primal or Dual problem. That is, it applies to the cases of infeasibility and unboundedness which we discussed in Chapter 3. Suppose the optimal solution to the Primal problem was unbounded. This would mean that the value of the objective function could be made infinitely large. But, the property we just presented states that the value of the objective function for the Dual problem must be greater than or equal to the value of the objective function for *any* feasible Primal solution. Therefore, we must conclude that if the Primal problem is unbounded, then the Dual problem will have no feasible solution. This property and its converse are presented below.

Property 5. If the solution to the Primal problem is unbounded, then the Dual problem has no feasible solution. Similarly, if the Dual problem is unbounded, then the Primal problem has no feasible solution.

Stated another way, Property 5 guarantees that the Dual problem will not have an optimal solution if the Primal problem does not have an optimal solution and vice versa. From Property 1, we know that if either of the problems has an optimal solution then so does the other, and that, in fact, both optimal solutions yield the same value for the objective function. Property 5 tells us what to expect when an optimal solution does not exist.

You should be cautioned that Property 5 does not rule out the possibility that both the Primal and Dual problems might be infeasible. In fact, examples can be constructed in which there is no feasible solution to either problem.

One other property of duality has not yet been discussed. Since this property is very important in sensitivity analysis, we defer discussion of it until later in the chapter.

4.4 Sensitivity Analysis

Sensitivity analysis is the study of how the optimal solution and the value of the optimal solution to a linear program change given changes in the various coefficients of the problem. That is, we are interested in answering questions such as the following: (1) what effect will a change in the coefficients in the objective function (c_j's) have?; (2) what effect will a change in the right-hand side values (b_i's) have?; and (3) what effect will a change in the coefficients in the constraining equations (a_{ij}'s) have? Since sensitivity analysis is concerned with how these changes affect the optimal solution, the analysis begins only after the optimal solution to the original linear programming problem has been obtained. Hence, sensitivity analysis can be referred to as postoptimality analysis.

There are several reasons why sensitivity analysis is considered so important from a managerial point of view. First, consider the fact that businesses operate in a dynamic environment. Prices of raw materials change over time; companies purchase new machinery to replace old; stock prices fluctuate; employee turnovers occur; etc. If a linear programming model has been used in a decision-making situation and later we find changes in the situation cause changes in some of the coefficients associated with the initial linear programming formulation, we would like to determine how these changes affect the optimal solution to our original linear programming problem. Sensitivity analysis provides us with this information without requiring us to completely solve a new linear program. For example, if the profit for the Par, Inc. Standard bags were reduced from \$10 to \$7 per bag, sensitivity analysis can tell the manager whether the production schedule of 540 Standard bags and 252 Deluxe bags is still the best decision or not. If it is, we will not have to solve a revised linear program with $7x_1 + 9x_2$ as the objective function.

Sensitivity analysis can also be used to determine how critical estimates of coefficients are in the solution to a linear programming problem. For example, suppose the management of Par, Inc. realizes the \$10 profit coefficient for Standard bags is a good, but rough, estimate of the profit the bags will actually provide. If sensitivity analysis shows Par, Inc. should produce 540 Standard bags and 252 Deluxe bags as long as the actual profit for Standard bags remains between $6.00 and $14.00, management can feel comfortable that the recommended production quantities are optimal. However, if the range for the profit of Standard bags is $9.90 to $12.00, management may want to reevaluate the accuracy of the $10.00 profit estimate. Management would especially want to consider what revisions would have to be made in the optimal production quantities if the profit for Standard bags dropped below the $9.90 limit.

As another phase of postoptimality analysis, management may want to investigate the possibility of adding resources to relax the binding constraints. In the Par, Inc. problem, management would possibly like to consider providing additional hours (e. g., overtime) for the cutting and dyeing and finishing operations. Sensitivity analysis can help answer the important questions of how much will each added hour be worth in terms of increasing profits and what is the maximum number of hours that can be added before a different basic solution becomes optimal.

Thus, you can see that through sensitivity analysis we will be able to provide additional valuable information for the decision maker. We begin our study of sensitivity analysis with the coefficients of the objective function.

4.5   Sensitivity Analysis—The Coefficients of the Objective Function

In this phase of sensitivity analysis, we will be interested in placing ranges on the values of the objective function coefficients such that as long as the actual value of the coefficient is within this range, the optimal solution will remain unchanged. As stated in the previous section, this information will tell us if we have to alter the optimal solution when a coefficient actually changes, and will provide us with an indication of how critical the estimates of the coefficients are in arriving at the optimal solution.

In the following sensitivity analysis procedures, we will be assuming that only one coefficient changes at a time and that all other objective function coefficients remain at the values defined in the initial linear programming model. To illustrate the analysis for the coefficients of the objective function, let us again consider the final Simplex tableau for the Par, Inc. problem.

BASIS	c_j	x_1 10	x_2 9	s_1 0	s_2 0	s_3 0	s_4 0	
x_2	9	0	1	30/16	0	−210/160	0	252
s_2	0	0	0	−15/16	1	25/160	0	120
x_1	10	1	0	−20/16	0	300/160	0	540
s_4	0	0	0	−11/32	0	45/320	1	18
	z_j	10	9	70/16	0	111/16	0	7668
	c_j-z_j	0	0	−70/16	0	−111/16	0	

Coefficients of the Nonbasic Variables

 The sensitivity analysis procedure for coefficients of the objective function depends upon whether we are considering the coefficient of a basic or nonbasic variable. For now, let us consider only the case of nonbasic variables.

 Since the nonbasic variables are not in solution, we are interested in the question of how much the objective function coefficient would have to change before it would be profitable to bring the associated variable into solution. Recall that it is only profitable to bring a variable into solution if its (c_j-z_j) entry in the net evaluation row is greater than or equal to zero.

 Suppose we denote a change in the objective function coefficient of variable x_j by Δc_j. Thus,

$$\Delta c_j = c_j' - c_j \qquad (4.1)$$

where

 c_j = the value of the coefficient of x_j in the original linear program

 c_j' = the new value of the coefficient of x_j.

Using this notation, we can write the new objective function coefficient as

$$c_j' = c_j + \Delta c_j \quad .$$
(4.2)

It will be desirable to bring the nonbasic variable, x_j, into solution if the new objective function coefficient is such that $c_j' - z_j > 0$ (i. e., if it will increase the value of the objective function). On the other hand, we will not want to bring the variable x_j into solution, and thus, will not change our current optimal solution as long as $c_j' - z_j \leq 0$. Our goal in this phase of sensitivity analysis is to determine the range of values c_j' can take on and still not affect the optimal solution.

Recall that z_j is computed by multiplying the coefficients of the <u>basic variables</u> (c_j column of the Simplex tableau) by the corresponding elements in the jth column of the $\overline{A}$ portion of the tableau. Thus, a change in the objective function coefficient for a nonbasic variable cannot affect the value of the z_j's. Therefore, the values of c_j' that do not require us to change the optimal solution are given by

$$c_j' - z_j \leq 0.$$

Since z_j will be known in the final Simplex tableau, any new coefficient c_j' for a nonbasic variable such that

$$c_j' \leq z_j$$

will not cause a change in the current optimal solution.

Note that there is no lower limit on the new coefficient c_j'. This is certainly as expected, since we have a maximization objective function and thus lower and lower c_j' values will make the nonbasic variables even less desirable.

Thus, for nonbasic variables, we can now establish a range of c_j' values which will not affect the current optimal solution. We call this range the *range of insignificance* for the nonbasic variables. It is given by:

$$- \infty < c_j' \leq z_j.$$

As long as the objective function coefficients for nonbasic variables remain within their respective ranges of insignificance, the nonbasic variables will remain at a zero value in the optimal solution. Thus, the current optimal solution and the value of the objective function at the optimal solution will not change.

Coefficients of the Basic Variables

Let us start by asking the question of how much would the objective function coefficient of a basic variable have to change before it would be profitable to change the current optimal solution. Again, realize that we will only change the current optimal solution if one or more of the net evaluation row values $(c_j - z_j)$ becomes greater than zero.

Let us consider a change in the objective function coefficient for the basic variable x_1 in the Par, Inc. problem. Let the new coefficient value be c_1'. Using equation (4.2), we can write $c_1' = c_1 + \triangle c_1$ where c_1 is the original coefficient 10 and $\triangle c_1$ is the change in the coefficient. Thus,

$$c_1' = 10 + \triangle c_1. \tag{4.3}$$

Let us now see what happens to the final Simplex tableau of the Par, Inc. problem where the objective function coefficient for x_1 becomes $10 + \triangle c_1$. This tableau is as follows:

BASIS	c_j	x_1 $10 + \triangle c_1$	x_2 9	s_1 0	s_2 0	s_3 0	s_4 0	
x_2	9	0	1	$30/16$	0	$-210/160$	0	252
s_2	0	0	0	$-15/16$	1	$25/160$	0	120
x_1	$10 + \triangle c_1$	1	0	$-20/16$	0	$300/160$	0	540
s_4	0	0	0	$-11/32$	0	$45/320$	1	18
z_j		$10 + \triangle c_1$	9	$\dfrac{70}{16} - \dfrac{20}{16}\triangle c_1$	0	$\dfrac{111}{16} + \dfrac{30}{16}\triangle c_1$	0	$7668 + 540\,\triangle c_1$
$c_j - z_j$		0	0	$-\dfrac{70}{16} + \dfrac{20}{16}\triangle c_1$	0	$-\dfrac{111}{16} - \dfrac{30}{16}\triangle c_1$	0	

How does the change of $\triangle c_1$ affect our final tableau? First, note that since x_1 is a basic variable, the new objective function coefficient $c_1' = 10 + \triangle c_1$ appears in the c_j column of the Simplex tableau. This means that the $10 + \triangle c_1$ value will affect the

z_j values for several of the variables. By looking at the z_j row, you can see that the new coefficient affects the z_j values of the basic variable x_1, both nonbasic variables (s_1 and s_3), and the objective function.

Recall that a decision to change the current optimal solution must be based on values in the net evaluation row. What variables have experienced a change in $(c_j - z_j)$ values because of the change Δc_1? As you can see, the change in the objective function coefficient for basic variable x_1 has caused changes in the $(c_j - z_j)$ values for both of the nonbasic variables. The $(c_j - z_j)$ values for all the basic variables remained unchanged; $(c_j - z_j) = 0$.

We have just identified the primary difference between the objective function sensitivity analysis procedures for basic and nonbasic variables. That is, a change in the objective function coefficient for a nonbasic variable only affects the $c_j - z_j$ value for that variable; however, a change in the objective function coefficient for a basic variable can affect the $c_j - z_j$ values for *all* nonbasic variables.

Returning to the Par, Inc. problem with the coefficient for x_1 changed to $10 + \Delta c_1$, we know that our current solution will remain optimal as long as all $(c_j - z_j) \leq 0$. Since the basic variables all still have $(c_j - z_j) = 0$, we will have to determine what range of values for Δc_1 will keep the $(c_j - z_j)$ values for all nonbasic variables less than or equal to zero.

For nonbasic variable s_1, we must have

$$- \;^{70}\!/_{16} + \;^{20}\!/_{16} \, \Delta c_1 \leq 0. \tag{4.4}$$

Solving for Δc_1, we see it will not be profitable to introduce s_1 as long as

$\quad \bullet \; ^{20}\!/_{16} \, \Delta c_1 \leq \;^{70}\!/_{16}$

$\qquad \Delta c_1 \leq \; ^{16}\!/_{20} \, (^{70}\!/_{16}) = \; ^{7}\!/_{2}$

$\qquad \Delta c_1 \leq 3.5.$

For nonbasic variable s_3, we must have

$$- \;^{111}\!/_{16} - \;^{30}\!/_{16} \, \Delta c_1 \leq 0. \tag{4.5}$$

Solving for $\triangle c_1$, we see it will not be profitable to introduce s_3 as long as

$$- \, {}^{30}\!\!/_{16} \, \triangle \, c_1 \leq \quad {}^{111}\!\!/_{16}$$

$$^{30}\!\!/_{16} \, \triangle \, c_1 \geq - \, {}^{111}\!\!/_{16}$$

$$\triangle \, c_1 \geq {}^{16}\!\!/_{30} \, (- \, {}^{111}\!\!/_{16}) = - \, {}^{111}\!\!/_{30}$$

$$\triangle \, c_1 \geq - \, 3.7.$$

Thus, in order to keep the net evaluation row values of the nonbasic variables less than or equal to zero, and keep the current optimal solution, changes in c_1 cannot exceed a 3.5 increase ($\triangle c_1 \leq 3.5$) or a 3.7 decrease ($\triangle c_1 \geq -3.7$). Hence, our current solution will remain optimal as long as:

$$-3.7 \leq \triangle c_1 \leq 3.5. \tag{4.6}$$

From equation (4.3), we know that $\triangle c_1 = c_1' - 10$ where c_1' is the new value of the coefficient for x_1 in the objective function. Thus we can use equation (4.6) to define a range for the coefficient values of x_1 that will not cause a change in the optimal solution. This is done as follows:

$$-3.7 \leq c_1' - 10 \leq 3.5;$$

therefore,

$$6.30 \leq c_1' \leq 13.50.$$

The above result indicates to the decision maker that as long as the profit for one Standard bag is between $6.30 and $13.50, the current production quantities of 540 Standard bags and 252 Deluxe bags will be optimal. We refer to the above range of values for the objective function coefficient of x_1 as the *range of optimality for c_1*.

To see how the management of Par, Inc. can make use of the above sensitivity analysis information, suppose that because of an increase in raw material prices, the profit of the Standard bag is reduced to $7 per unit. The range of optimality for c_1 indicates that the current solution $x_1 = 540$, $x_2 = 252$, $s_1 = 0$, $s_2 = 120$, $s_3 = 0$, and $s_4 = 18$ will still be optimal. To see the effect of this change, let us calculate the final Simplex tableau for the Par, Inc. problem after c_1 has been reduced to \$7.

BASIS	c_j	x_1 7	x_2 9	s_1 0	s_2 0	s_3 0	s_4 0	
x_2	9	0	1	30/16	0	−210/160	0	252
s_2	0	0	0	−15/16	1	25/160	0	120
x_1	7	1	0	−20/16	0	300/160	0	540
s_4	0	0	0	−11/32	0	45/320	1	18
z_j		7	9	130/16	0	21/16	0	6048
c_j-z_j		0	0	−130/16	0	−21/16	0	

Since all of the $(c_j - z_j)$ values are less than or equal to zero, the solution is optimal. As you can see, this solution is the same as our previous optimal solution. Note, however, that because of the decrease in profit for the Standard bags, the total profit has been reduced to $6048 = 7668 + 540 \triangle c_1 = 7668 + 540(- 3)$.

What would happen if the profit per Standard bag were reduced to \$5? Again, we refer to the range of optimality for c_1. Since $c_1 = 5$ is outside the range, we know that a change this large will cause a new solution to be optimal. Consider the following Simplex tableau containing the same basic feasible solution but with the value of $c_1 = 5$.

BASIS	c_j	x_1 5	x_2 9	s_1 0	s_2 0	s_3 0	s_4 0	
x_2	9	0	1	30/16	0	−210/160	0	252
s_2	0	0	0	−15/16	1	25/160	0	120
x_1	5	1	0	−20/16	0	300/160	0	540
s_4	0	0	0	−11/32	0	45/320	1	18
z_j		5	9	170/16	0	−39/16	0	4968
c_j-z_j		0	0	−170/16	0	39/16	0	

As expected, the solution, $x_1 = 540$, $x_2 = 252$, $s_1 = 0$, $s_2 = 120$, $s_3 = 0$, and $s_4 = 18$, is no longer optimal! The coefficient for s_3 in the net evaluation row is now greater than zero. This implies that at least one more iteration must be performed to reach the optimal solution. Check for yourself to see that the new optimal solution will require production of 300 Standard bags and 420 Deluxe bags.

Thus, we see how the range of optimality can be used to quickly determine whether or not a change in the objective function coefficient of a basic variable will cause a change in the optimal solution. Note that by using the range of optimality to determine whether or not the change in a profit coefficient for a basic variable is large enough to cause a change in the optimal solution, we can avoid the time-consuming process of reformulating and resolving the entire linear programming problem.

The general procedure for determining the range of optimality for the basic variable associated with column j and row i of the Simplex tableau is to first find the range of values for $\triangle c_j$ that satisfy

$$(c_k - z_k) - \bar{a}_{ik} \triangle c_j \le 0 \qquad (4.7)$$

for each nonbasic variable x_k or s_k. In this process, we will obtain a limit on $\triangle c_j$ for each nonbasic variable. The most restrictive upper and lower limits on $\triangle c_j$ will be used to define the range of optimality. For example, if one of these inequalities requires $\triangle c_j \le 3$ and another requires $\triangle c_j \le 1$, the $\triangle c_j \le 1$ is the most restrictive limit and will be the upper limit for $\triangle c_j$.

After considering every nonbasic variable, we will have found upper and lower limits on $\triangle c_j$ in the following form:

$$\alpha \le \triangle c_j \le \beta$$

where

$\alpha =$ lower limit,

$\beta =$ upper limit.[3]

---

[3] α is $-\infty$ if inequalities (4.7) do not provide a lower limit on $\triangle c_j$ and β is $+\infty$ if inequalities (4.7) do not provide an upper limit.

Using the relationship $(\Delta c_j = c_j' - c_j)$, the range of optimality is then given by

$$\alpha \leq c_j' - c_j \leq \beta$$

or

$$c_j + \alpha \leq c_j' \leq c_j + \beta \tag{4.8}$$

Applying the above procedure to the objective function coefficient c_2 in our Par, Inc. problem (the profit per unit for Deluxe bags), we see that in order to satisfy inequalities (4.7), Δc_2 must satisfy

$$(-\,^{7}\!\%_{16}) - (^{30}\!\%_{16})\,\Delta\,c_2 \leq 0 \tag{4.9}$$

and

$$(-\,^{111}\!\%_{16}) - (-\,^{21}\!\%_{16})\,\dot{\Delta}\,c_2 \leq 0. \tag{4.10}$$

From (4.9), we get

$$\Delta\,c_2 \geq -\,\%_3.$$

And from (4.10), we get

$$\Delta\,c_2 \leq {}^{111}\!\%_{21}.$$

Thus we have

$$-\,\%_3 \leq \Delta\,c_2 \leq {}^{111}\!\%_{21}.$$

Using equation (4.8) and our original profit coefficient of $c_2 = 9$, we have the following range of optimality for c_2:

$$9 - \%_3 \leq c_2' \leq 9 + {}^{111}\!\%_{21}$$

or

$$6.67 \leq c_2' \leq 14.29.$$

Thus, we see that as long as the profit of Deluxe bags is between \$6.67 per unit and \$14.29 per unit, the production quantities of 540 Standard bags and 252 Deluxe bags will remain optimal.

As a summary, we present the following managerial interpretation of sensitivity analysis for the objective function coefficients. Think of the basic variables as corresponding to our current product line and the nonbasic variables as representing other products we might produce. Within bounds, changes in the profit associated with one of the products in our current product line would not cause us to change our product mix or the amounts produced, but the changes would have an effect on our total profit. Of course, if the profit associated with one of our products changed drastically, we would change our product line (i. e., move to a different basic solution). For products we are not currently producing (nonbasic variables), it is obvious that a decrease in per unit profit would not make us want to produce them; however, if the per unit profit for one of these products became large enough, we would want to consider adding it to our product line.

4.6 Sensitivity Analysis—The Right-Hand Sides

A very important phase of sensitivity analysis, both from a theoretical and practical point of view, is the study of the effect of changes of the right-hand sides on the optimal solution and the value of the optimal solution. By changes of the right-hand sides, we mean simply changing the values of one of the elements in the b vector in the matrix representation of a linear program. We shall see that the Dual plays a major role in this analysis.

Quite often in linear programming problems we can interpret the b_i's as the resources available. For example, in the Par, Inc. problem, the right-hand side values represented the number of man-hours available in each of four departments. Thus, valuable management information could be provided if we knew how much it would be worth to the company if one or more of these production time resources were increased. Sensitivity analysis of the right-hand sides can help provide this information.

The Dual Variables as Shadow Prices

As we saw earlier, in our study of the relationships between the Primal and the Dual problems, there is a dual variable associated with every constraint of the Primal problem. It turns out that the values of these dual variables indicate how much the value of the optimal solution to the Primal problem will change given a change in one of the right-hand side values. This is another property of the relationships between the Primal and Dual problems.

Property 6. If b_i (the right-hand side of the ith primal constraint) is increased by one unit, the value of the optimal solution to the Primal problem will change by u_i units, where u_i is the value of the dual variable corresponding to the ith primal constraint.[4]

From Property 2 of the Primal-Dual relationship, we know that we can read the values of the optimal dual variables for the Par, Inc. problem from the final Simplex tableau of the Primal problem.

		x_1	x_2	s_1	s_2	s_3	s_4	
BASIS	c_j	10	9	0	0	0	0	
x_2	9	0	1	30/16	0	−210/160	0	252
s_2	0	0	0	−15/16	1	25/160	0	120
x_1	10	1	0	−20/16	0	300/160	0	540
s_4	0	0	0	−11/32	0	45/320	1	18
	z_j	10	9	70/16	0	111/16	0	7668
	c_j-z_j	0	0	−70/16	0	−111/16	0	

Recall that the values of the dual variables are given by the negative of the $(c_j - z_j)$ entries for the slack variables in the Simplex tableau corresponding to the optimal solution of the Primal problem. Thus, you can see from the above tableau that the values of the dual variables are $u_1 = 70/16$, $u_2 = 0$, $u_3 = 111/16$, and $u_4 = 0$.

According to Property 6, an increase in b_2 or b_4 of one unit should have no effect on the objective function. Clearly this is true for the Par, Inc. problem since the sewing and inspection and packaging constraints are satisfied as strict *inequalities*. That is, the slack variable associated with each of these constraints is in the basic solution. Since there is slack time in

---

[4] We assume here that the same basis will remain optimal after making this increase in b_i. The reason for this assumption will be made clear later.

both of these departments, we could not expect the value of the optimal solution to change when we increased the amount of slack time. In other words, we already have too much of each of these resources.

Notice that the dual variables corresponding to the first and third constraints are positive. This is to be expected since both of these constraints are binding in the optimal solution. Thus, if we could increase the amount of cutting and dyeing and/or finishing time available, we could expect the value of the optimal solution to increase. The dual variable $u_1 = \frac{70}{16}$ indicates that the value of the optimal solution will increase by $\frac{70}{16}$ if b_1—the maximum available cutting and dyeing time—is increased by one unit. Similarly, the dual variable $u_3 = \frac{111}{16}$ indicates that the value of the optimal solution will increase by $\frac{111}{16}$ if b_3—the maximum available finishing time—is increased by one unit.

Let us consider the cutting and dyeing operation in detail. Suppose we increase the original 630 hours available to 631 hours and then compute the revised final Simplex tableau. From this tableau, which is shown below, we see that the objective function has increased by $\frac{70}{16}$ as expected.

BASIS	c_j	x_1 10	x_2 9	s_1 0	s_2 0	s_3 0	s_4 0	
x_2	9	0	1	$30/16$	0	$-210/160$	0	$253\frac{14}{16}$
s_2	0	0	0	$-15/16$	1	$25/160$	0	$119\frac{1}{16}$
x_1	10	1	0	$-20/16$	0	$300/160$	0	$538\frac{12}{16}$
s_4	0	0	0	$-11/32$	0	$45/320$	1	$17\frac{21}{32}$
z_j		10	9	$70/16$	0	$111/16$	0	$7672\frac{6}{16}$
$c_j - z_j$		0	0	$-70/16$	0	$-111/16$	0	

You are probably wondering how this final tableau was computed. Certainly, we did not go through all of the Simplex iterations again after changing b_1 from 630 to 631. You can see that the only changes in the tableau are the differences in the values of the basic variables (i. e., the last column). The entries in this last column of the Simplex tableau have been obtained by merely adding the first five entries in the third column to the last column in the previous tableau. That is,

$$
\begin{bmatrix}
30/16 \\
-15/16 \\
-20/16 \\
-11/32 \\
70/16
\end{bmatrix}
+
\begin{bmatrix}
252 \\
120 \\
540 \\
18 \\
7668
\end{bmatrix}
=
\begin{bmatrix}
253^{14}/_{16} \\
119^{1}/_{16} \\
538^{12}/_{16} \\
17^{21}/_{32} \\
7672^{6}/_{16}
\end{bmatrix}
$$

The reason for this is as follows. The entries in column s_1 tell us how many units of the basic variables will be driven out of solution if one unit of variable s_1 is introduced into solution. Increasing b_1 by one unit has just the reverse effect. That is, it is just the same as taking one unit of s_1 out of solution. If b_1 is increased by one unit, then the entries in the s_1 column tell us how many units of each of the basic variables may be added to solution for every unit of b_1 added. The increase in the value of the objective function corresponding to adding one unit of b_1 is equal to the negative of $(c_3 - z_3)$, or equivalently, the value of the dual variable u_1.

Since this is such an important property, let us state it again. That is, associated with every constraint of the Primal problem is a dual variable. The value of the dual variable indicates how much the objective function of the Primal problem will increase as the value of the right-hand side of the associated primal constraint is increased by one unit.

In other words, the value of the dual variable indicates the value of one additional unit of a particular resource. Hence, this value can be interpreted as the maximum value or price we would be willing to pay to obtain the one additional unit of the resource. Because of this interpretation, the value of one additional unit of a resource is often called the *shadow price* of the resource. Thus, the values of the dual variables are often referred to as shadow prices.

Range of Feasibility for the Right-Hand Side Values

The dual variables can be used to predict the change in the value of the objective function corresponding to a unit change in one of the b_i's. But here is the "catch". The dual variables can be used in this fashion only as long as the change in the b_i's is not large enough to make the current basis infeasible.

Thus, we will be interested in determining how much a particular b_i can be changed without causing a change in the current optimal basis. In effect we will do this by calculating a range of values over which a particular b_i can vary without any of the current basic variables becoming infeasible. This range of values will be referred to as the *range of feasibility*.

To demonstrate the effect of increasing a resource by several units, consider increasing the available cutting and dyeing time for the Par, Inc. problem by 10 hours. Will the new basic solution be feasible? If so, we can expect an increase in the objective function of $10u_1 = 10(7\%_{16}) = \$43.75$. As before we can calculate the new values of the basic variables by adding to the old values the change in b_1 times the coefficients in the s_1 column.

$$\text{New solution} = \begin{bmatrix} 252 \\ 120 \\ 540 \\ 18 \end{bmatrix} + 10 \begin{bmatrix} 30/_{16} \\ -15/_{16} \\ -20/_{16} \\ -11/_{32} \end{bmatrix} = \begin{bmatrix} 270^{12}/_{16} \\ 110^{10}/_{16} \\ 527\%_{16} \\ 14\%_{16} \end{bmatrix}$$

Since the new solution is still feasible (i. e., all of the basic variables are ≥ 0), the prediction made by the dual variable of a \$43.75 change in the objective function resulting from a 10-unit increase in b_1 is correct.

How do we know when a change in b_1 is so large that the current basis will become infeasible? We shall first answer this question specifically for the Par, Inc. problem. But, you should realize that the following procedure applies only for the less-than-or-equal-to constraints of a linear program. The procedures for the cases of greater-than-or-equal-to and equality constraints are discussed later in this section and in problem 22 at the end of the chapter.

We begin by showing how to calculate upper and lower bounds for the maximum amount b_1 can be changed before the current basis becomes infeasible. Given a change in b_1 of $\Delta\, b_1$, the *new basic solution* is given by

$$\text{New solution} = \begin{bmatrix} x_2 \\ s_2 \\ x_1 \\ s_4 \end{bmatrix} = \begin{bmatrix} 252 \\ 120 \\ 540 \\ 18 \end{bmatrix} + \Delta\, b_1 \begin{bmatrix} {}^{30}\!/_{16} \\ -{}^{15}\!/_{16} \\ -{}^{20}\!/_{16} \\ -{}^{11}\!/_{32} \end{bmatrix} = \begin{bmatrix} 252 + {}^{30}\!/_{16}\, \Delta\, b_1 \\ 120 - {}^{15}\!/_{16}\, \Delta\, b_1 \\ 540 - {}^{20}\!/_{16}\, \Delta\, b_1 \\ 18 - {}^{11}\!/_{32}\, \Delta\, b_1 \end{bmatrix} \qquad (4.11)$$

As long as the value of each variable in the new basic solution remains nonnegative, the new basic solution will remain feasible and, therefore, optimal. We can keep the variables nonnegative by limiting the change in b_1 (i. e., $\Delta\, b_1$) so that we satisfy each of the following conditions:

$$252 + {}^{30}\!/_{16}\, \Delta\, b_1 \geq 0 \qquad\qquad\qquad (4.12)$$
$$120 - {}^{15}\!/_{16}\, \Delta\, b_1 \geq 0 \qquad\qquad\qquad (4.13)$$
$$540 - {}^{20}\!/_{16}\, \Delta\, b_1 \geq 0 \qquad\qquad\qquad (4.14)$$
$$18 - {}^{11}\!/_{32}\, \Delta\, b_1 \geq 0. \qquad\qquad\qquad (4.15)$$

Note that the left-hand sides of the above inequalities represent the values of the basic variables after b_1 has been changed by $\Delta\, b_1$.

Working algebraically with the four inequalities ((4.12), (4.13), (4.14), and (4.15)), we get

$$\Delta\, b_1 \geq (\ {}^{16}\!/_{30})\, (-252) = -134.4$$
$$\Delta\, b_1 \leq (-{}^{16}\!/_{15})\, (-120) = 128$$
$$\Delta\, b_1 \leq (-{}^{16}\!/_{20})\, (-540) = 432$$
$$\Delta\, b_1 \leq (-{}^{32}\!/_{11})\, (-\ 18) = 52{}^{4}\!/_{11}.$$

Since we must satisfy all four inequalities, the most restrictive limits on $\Delta\, b_1$ must be used; therefore, we have

$$-134.4 \leq \Delta\, b_1 \leq 52{}^{4}\!/_{11}.$$

Using $\Delta\, b_1 = b_1' - b_1$, where b_1' is the new number of hours available in the cutting and dyeing department and b_1 is the original number of hours available ($b_1 = 630$), we have the following

$$-134.4 \leq b_1' - 630 \leq 52\tfrac{8}{11}$$

or

$$495.6 \leq b_1' \leq 682\tfrac{8}{11}.$$

The above range of values for b_1' indicates that as long as the time available in the cutting and dyeing department is between 495.6 and $682\tfrac{8}{11}$ hours, the current basis will remain feasible and optimal. Thus, this is the range of feasibility for the right-hand side value of the cutting and dyeing constraint.

Since $u_1 = \tfrac{70}{16}$, we know we can improve profit by $\tfrac{70}{16}$ for an additional hour of cutting and dyeing time. Suppose we increase b_1 by $52\tfrac{8}{11}$ hours to the upper limit of its range of feasibility, $682\tfrac{8}{11}$. The new profit with this change will be $7668 + (\tfrac{70}{16})(52\tfrac{8}{11}) = 7897\tfrac{1}{11}$, and the new optimal solution (using (4.11)) is

$$x_2 = 252 + \tfrac{30}{16}(52\tfrac{8}{11}) = 350\tfrac{2}{11}$$
$$s_2 = 120 - \tfrac{15}{16}(52\tfrac{8}{11}) = 70\tfrac{10}{11}$$
$$x_1 = 540 - \tfrac{20}{16}(52\tfrac{8}{11}) = 474\tfrac{6}{11}$$
$$s_4 = 18 - \tfrac{11}{32}(52\tfrac{8}{11}) = 0$$

with the nonbasic variables s_1 and s_3 still equal to zero.

What has happened to our solution in the above process? You can see that the increased cutting and dyeing time has caused us to revise the production plan so that Par will produce a greater number of Deluxe bags and a slightly smaller number of Standard bags. Overall, the profit will be increased by $(\tfrac{70}{16})(52\tfrac{8}{11}) = \$229\tfrac{1}{11}$.

Problem 16 at the end of the chapter will ask you to show graphically what happens as we increase the cutting and dyeing time resource to its upper limit of $682\tfrac{8}{11}$ hours.

Our procedure for determining the range of feasibility has involved only the cutting and dyeing constraint. The procedure for calculating the range of feasibility for the right-hand side value of any less-than-or-equal-to constraint is the same. The first step (paralleling (4.11)) for a general constraint i is to calculate the range of values for $\triangle b_i$ that satisfy the conditions shown below.

$$\begin{bmatrix} \bar{b}_1 \\ \bar{b}_2 \\ \cdot \\ \cdot \\ \cdot \\ \cdot \\ \bar{b}_m \end{bmatrix} + \triangle b_i \begin{bmatrix} \bar{a}_{1j} \\ \bar{a}_{2j} \\ \cdot \\ \cdot \\ \cdot \\ \cdot \\ \bar{a}_{mj} \end{bmatrix} \geq \begin{bmatrix} 0 \\ 0 \\ \cdot \\ \cdot \\ \cdot \\ \cdot \\ 0 \end{bmatrix} \tag{4.16}$$

Current solution (last column of the final Simplex tableau)

Column of the final Simplex tableau corresponding to the slack variable associated with constraint i

This determines a lower and an upper limit on $\triangle b_i$. The range of feasibility can then be established.

Similar arguments to the ones presented in this section can be used to develop a procedure for determining the range of feasibility for the right-hand side value of a greater-than-or-equal-to constraint. Essentially, the procedure is the same with the column corresponding to the surplus variable associated with the constraint playing the central role. For a general greater-than-or-equal-to constraint i, we first calculate the range of values for $\triangle b_i$ that satisfy the conditions shown in (4.17).

$$\begin{bmatrix} \bar{b}_1 \\ \bar{b}_2 \\ \cdot \\ \cdot \\ \cdot \\ \cdot \\ \bar{b}_m \end{bmatrix} - \triangle b_i \begin{bmatrix} \bar{a}_{1j} \\ \bar{a}_{2j} \\ \cdot \\ \cdot \\ \cdot \\ \cdot \\ \bar{a}_{mj} \end{bmatrix} \geq \begin{bmatrix} 0 \\ 0 \\ \cdot \\ \cdot \\ \cdot \\ \cdot \\ 0 \end{bmatrix} \tag{4.17}$$

Current solution

Column of the final Simplex tableau corresponding to the surplus variable associated with constraint i

Once again, these conditions establish a lower and an upper limit on $\triangle b_i$. Given these limits, the range of feasibility is easily determined.

To calculate the range of feasibility for the right-hand side value of an equality constraint, we use the column of the Simplex tableau corresponding to the artificial variable associated with that constraint. The limits on $\triangle b_i$ are given by the $\triangle b_i$ values satisfying (4.18).

$$
\begin{bmatrix} \bar{b}_1 \\ \bar{b}_2 \\ \cdot \\ \cdot \\ \cdot \\ \cdot \\ \bar{b}_m \end{bmatrix} + \triangle b_i \begin{bmatrix} \bar{a}_{1j} \\ \bar{a}_{2j} \\ \cdot \\ \cdot \\ \cdot \\ \cdot \\ \bar{a}_{mj} \end{bmatrix} \geq \begin{bmatrix} 0 \\ 0 \\ \cdot \\ \cdot \\ \cdot \\ \cdot \\ 0 \end{bmatrix} \tag{4.18}
$$

Current solution

Column of the final Simplex tableau corresponding to the artificial variable associated with constraint i

As long as the change in a b_i is such that the value of b_i' stays within its range of feasibility, the same basis will remain feasible and optimal. Changes that force b_i' outside its range of feasibility will force us to perform additional Simplex iterations to find the new optimal basic feasible solution. More advanced linear programming texts show how this can be done without completely resolving the problem. In any case, the calculation of the range of feasibility for each of the b_i's is valuable management information and should be included as part of the management report on any linear programming project.

Note that the procedure for determining the range of feasibility for the right-hand side values has involved changing only *one* value at a time (i. e., only one $\triangle b_i$). Multiple changes are more difficult to analyze and in practice usually require the complete generation of a new final Simplex tableau.

4.7   Sensitivity Analysis—The A Matrix

A change in one of the coefficients of the A matrix can have a significant effect on the optimal solution to a linear programming problem. A complete discussion of the ramifications of making changes in the A matrix is beyond the scope of this text; however, we can make a few remarks based on an analysis of the graphical solution of the example problem shown below:

$$\max\ 2x_1 + 3x_2$$

s.t.

$$1x_1 + 2x_2 \leq\ 8$$
$$3x_1 + 2x_2 \leq\ 12$$
$$x_1,\ x_2 \geq 0.$$

The graphical solution to this problem is shown in Figure 4–A.

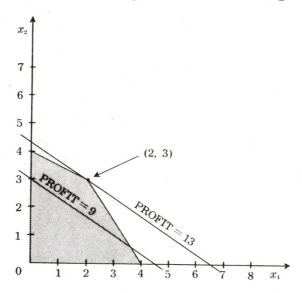

Figure 4–A.   Graphical Solution of the Example Problem

The A matrix for the problem is

$$A = \begin{bmatrix} 1 & 2 \\ 3 & 2 \end{bmatrix}$$

Suppose we make a change in the A matrix such that $a_{12} = 3$. The problem now becomes

max $2x_1 + 3x_2$

s.t.

$$1x_1 + 3x_2 \leq 8$$
$$3x_1 + 2x_2 \leq 12$$
$$x_1, x_2 \geq 0.$$

The graphical solution to this problem is shown in Figure 4–B.

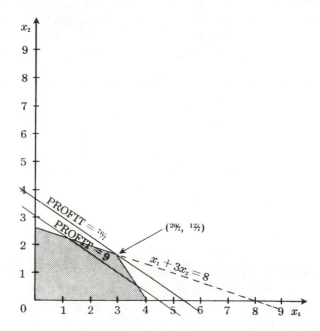

Figure 4–B.   Graphical Solution of the Example Problem After Changing a_{12} from 2 to 3

The optimal solution of $x_1 = 2$, $x_2 = 3$, and $z = 13$ for the original example has been revised to $x_1 = {}^{20}\!/_7$, $x_2 = {}^{12}\!/_7$, and $z = {}^{76}\!/_7$ as a result of the change in the A matrix.   Thus, the effect of this change in A was to reduce the amount of x_2 in the optimal solution and to increase the amount of x_1.   Graphically, we see that the slope and the x_2 intercept of the first constraint have changed.

This change has altered the feasible region and has caused a change in the optimal solution.

The way to analyze, qualitatively, the effects of a change in a_{ij} for basic variable x_j is to think about the effect this change has on the constraining equation. Let us assume for the moment that all of our constraints are of the less-than-or-equal-to type. Then, if a_{ij} is increased and constraint i is binding, the value of the objective function will decrease, and the amount of x_j in the optimal solution will also decrease (assuming $x_j > 0$; i. e., no degeneracy). If a_{ij} is decreased and constraint i is binding, then the value of the optimal solution will increase, and the amount of x_j will increase in the optimal solution. The converse of the above is true if our constraints are of the greater-than-or-equal-to type. For the case of equality constraints, additional analysis is necessary to determine the effect on the optimal solution. If one of the a_{ij} should change for a nonbinding constraint, this will have no effect on the optimal solution unless the change is large enough to make a new basis optimal. In addition, if one of the a_{ij}'s change for a nonbasic variable, the change will have no effect on the optimal solution unless the change is large enough to cause the $(c_j - z_j)$ for that variable to become greater than or equal to zero.

A range can be computed for any a_{ij} such that values within the range do not require a new basis for the optimal solution. However, the calculations are much more complicated than they were for the coefficients of the objective function and for the right-hand sides. Thus, a detailed analysis of changes in the A matrix is beyond the scope of this text.

4.8 Summary

The purpose of this chapter was to introduce the reader to the two related topics of duality and sensitivity analysis. We began our study by defining what was meant by a Primal linear programming problem and a Dual linear programming problem. The Primal problem in matrix notation was

$$\max \ c^t x$$

s.t.

$$Ax \le b$$
$$x \ge 0.$$

The Dual problem was given as

$$\min \, b^t u$$

s.t.

$$A^t u \geq c$$

$$u \geq 0.$$

We saw that in solving either problem we had in effect solved the other problem. Two of the important properties of duality are restated below:

Property 1. If the Dual problem has an optimal solution, the Primal problem has an optimal solution and vice versa. Furthermore, the values of the optimal solutions to the Dual and Primal problems are equal.

Property 2. The optimal values of the dual variables are given by the negative of the $(c_j - z_j)$ entries for the slack variables in the Simplex tableau corresponding to the optimal Primal solution. Similarly, the optimal values of the primal variables are given by the negative of the $(c_j - z_j)$ entries for the surplus variables in the Simplex tableau corresponding to the optimal Dual solution.

Another important duality property was found to be useful in performing sensitivity analysis for the right-hand side values. This was Property 6.

Property 6. If b_i (the right-hand side value for the ith primal constraint) is increased by one unit, the value of the optimal solution to the Primal problem will change by u_i units, where u_i is the value of the dual variable corresponding to the ith primal constraint.

This property allowed us to predict that the change in the value of the objective function resulting from a change in b_i of Δb_i would be $u_i \Delta b_i$ provided that the optimal basis did not change.

We showed, through sensitivity analysis, how we could provide the decision maker with valuable information as to how the optimal solution and value of that solution could be expected to change given a change in one of the coefficients of the problem. With respect to the objective function coefficients, we calculated a range of optimality over which the basic variables could range with no resultant change in the optimal solution. Similarly, a range of insignificance was calculated for the objective function coefficients of the nonbasic variables. For the right-hand side values, we computed a range of feasibility over which the b_i's could vary without causing the current basis to become infeasible. With respect to the A matrix, our remarks were qualitative since calculating a range of feasibility for the a_{ij}'s is outside the scope of this text.

The calculations and the resulting conclusions we obtained from sensitivity analysis were based on the assumption that *only one* coefficient was being changed at a time and that all other coefficients remained fixed. It is possible through sensitivity analysis to study the effects of multiple changes in the problem coefficients, but the calculations are more complicated. Hence, we leave this topic to the more advanced linear programming texts.

4.9 Glossary

1. *Primal problem*—As defined in this text, a linear program which consists of a maximization objective function subject to less-than-or-equal-to constraints. In matrix notation, it is

$$\max\ c^t x$$
$$\text{s.t.}$$
$$Ax \leq b$$
$$x \geq 0.$$

2. *Dual problem*—This is the Dual of a Primal problem. It is a linear program which consists of a minimization objective function subject to greater-than-or-equal-to constraints.

In matrix notation, it is

$$\min\ b^t u$$
$$\text{s.t.}$$
$$A^t u \geq c$$
$$u \geq 0.$$

3. *Complementary slackness*—A duality property which states that a strictly positive (> 0) dual variable in the optimal dual solution indicates that the corresponding primal constraint is satisfied as an equality (i. e., binding) in the optimal solution to the Primal problem.

4. *Shadow price*—The value of a dual variable. It indicates the value of one additional unit of the resource associated with the corresponding primal constraint.

5. *Range of optimality*—The range of values over which a c_j associated with a basic variable may vary without causing any change in the optimal solution (i. e., the values of all the variables will remain the same, but the value of the objective function will change).

6. *Range of insignificance*—The range of values over which a c_j associated with a nonbasic variable may vary without causing any change in the optimal solution or the value of the objective function.

7. *Range of feasibility*—The range of values over which a b_i may vary without causing the current basic solution to become infeasible. The values of the variables in solution will change, but the same variables will remain basic.

4.10 Problems

1. Suppose that the following problem is a product-mix problem with x_1, x_2, x_3, and x_4 indicating the units of product 1, 2, 3, and 4 respectively.

   $$\max 4x_1 + 6x_2 + 3x_3 + x_4$$

   s.t.

   $$1.5x_1 + 2x_2 + 4x_3 + 3x_4 \leq 550 \quad \text{(Machine A Hrs.)}$$
   $$4x_1 + x_2 + 2x_3 + x_4 \leq 700 \quad \text{(Machine B Hrs.)}$$
   $$2x_1 + 3x_2 + x_3 + 2x_4 \leq 200 \quad \text{(Machine C Hrs.)}$$
   $$x_1, x_2, x_3, x_4 \geq 0.$$

   a. Formulate the Dual to this problem.

   b. Solve the Dual. Use the dual solution to show the profit-maximizing product-mix is $x_1 = 0$, $x_2 = 25$, $x_3 = 125$, and $x_4 = 0$.

c. Use the dual variables to identify the machine or machines that are producing at maximum capacity. If the manager can select one machine for additional production capacity which machine should have priority? Why?

2. Find the Dual problem for the linear program given below.

max $10x_1 + 9x_2 + 4x_3 + 6x_4$

s.t.

$$3x_1 + 2x_2 + 4x_3 + 2x_4 \leq 70$$
$$5x_1 + 5x_2 + \ x_3 + 3x_4 \leq 60$$
$$5x_1 + 6x_2 + 3x_3 + \ x_4 \leq 25$$
$$x_1, x_2, x_3, x_4 \geq 0.$$

3. Write the following linear program as a Primal problem.

max $5x_1 + \ x_2 + 3x_3$

s.t.

$$x_1 + \ x_2 \qquad \geq 40$$
$$2x_1 + 3x_2 + \ x_3 \leq 50$$
$$3x_1 + 2x_2 + 2x_3 \leq 25$$
$$x_2 + \ x_3 \geq 10$$
$$x_1, x_2, x_3 \geq 0$$

From a computational point of view would you rather solve the above linear programming problem or the Dual problem?

4. Find the Primal problem for the linear program given below.

max $3x_1 + \ x_2 + 5x_3 + 3x_4$

s.t.

$$3x_1 + \ x_2 + 2x_3 \qquad = 30$$
$$2x_1 + \ x_2 + 3x_3 + \ x_4 \geq 15$$
$$2x_2 \qquad + 3x_4 \leq 25$$
$$x_1, x_2, x_3, x_4 \geq 0$$

5. Write the Dual of problem 4.

6. Consider the following linear program

max $2x_1 + 3x_2$

s.t.

$$x_1 + 2x_2 \leq 8$$
$$3x_1 + 2x_2 \leq 12$$
$$x_1, x_2 \geq 0.$$

   a. Write the Dual of this problem.

   b. Solve both the Primal and Dual problems by the graphical procedure.

   c. Solve both the Primal and Dual problems by the Simplex method.

   d. Using your results from parts b) and c), identify where and how you can observe properties 1, 2, 3, and 4 of the Primal-Dual relationship.

7. The Photo Chemicals' problem presented in Chapter 3 asked you to determine the minimum cost production plan for two of its products. The initial formulation was as follows:

min $x_1 + x_2$

s.t.

$$x_1 \qquad\quad \geq 30 \quad \text{Min. Product 1}$$
$$x_2 \geq 20 \quad \text{Min. Product 2}$$
$$x_1 + 2x_2 \geq 80 \quad \text{Min. Raw Material}$$
$$x_1, x_2 \geq 0$$

   a. Write this linear program as a Primal problem.

   b. Show the Dual problem.

   c. Solve the Dual problem and show that the optimal production plan is $x_1 = 30$ and $x_2 = 25$.

   d. Recall that the third constraint involved a management request that the current 80 pounds of a perishable raw material be used as soon as possible. However, after learning that the optimal solution calls for an excess production of five units of product 2, management is reconsidering the raw material requirement. Specifically, you have been asked to identify the cost effect if this constraint were relaxed. Use the dual variable to indicate the change in the cost if only 79 pounds of raw material had to be used.

8. Write the Dual problem for the following linear program.

min $3x_1 + x_2 + 2x_3$

s.t.

$$2x_1 + x_2 + 3x_3 = 5$$
$$4x_1 + x_2 + x_3 = 4$$
$$2x_1 \quad + x_3 \leq 7$$
$$x_1 + 2x_2 \quad \geq 4$$
$$x_1, x_2, x_3 \geq 0$$

9. Find the Dual problem for the following linear programming problem.

min $4x_1 + 3x_2 + 6x_3$

s.t.

$$x_1 + .5x_2 + x_3 \geq 15$$
$$2x_2 + x_3 \geq 30$$
$$x_1 + x_2 + 2x_3 \geq 20$$
$$x_1, x_2, x_3 \geq 0$$

10. A salesman who sells two products is trying to determine the number of calls he should make during the next month to promote each product. Based on past experience, he averages a $10 commission for every call for product 1 and a $5 commission for every call for product 2. The company requires at least 20 calls per month for each product and not more than 100 calls per month on any one product. In addition, the salesman spends about 3 hours for each call for product 1 and 1 hour for each call for product 2. If he has a total of 175 selling hours available next month, how many calls should be made for each of the two products in order to maximize his commission.

a. Formulate the problem.

b. Since we have five constraints and only two variables, solve the Dual problem.

c. Find the optimal solution and interpret all the dual variables.

11. Consider the linear program given below.

$$\max 15x_1 + 30x_2 + 20x_3$$

s.t.

$$x_1 \qquad\quad + \quad x_3 \le 4$$
$$\tfrac{1}{2}x_1 + 2x_2 + \quad x_3 \le 3$$
$$x_1 + \quad x_2 + 2x_3 \le 6$$
$$x_1,\, x_2,\, x_3,\, \ge 0$$

a. Find the optimal solution.

b. Calculate the range of optimality or range of insignificance (whichever is appropriate) for c_1.

c. What would be the effect of a five unit increase in c_1 (from 15 to 20) on the optimal solution and the value of that solution?

d. Calculate the range of optimality or range of insignificance (whichever is appropriate) for c_3.

e. What would be the effect of a five unit increase in c_3 (from 20 to 25) on the optimal solution and the value of that solution?

12. Consider again the linear programming problem presented in problem 11.

a. What are the values of the dual variables?

b. How much will the value of the objective function change if b_1 is increased from 4 to 5?

c. How much will the value of the objective function change if b_2 is increased from 3 to 4?

d. How much will the value of the objective function change if b_3 is increased from 6 to 7?

13. Consider the following linear program.

$$\max 3x_1 + \quad x_2 + 5x_3 + 3x_4$$

s.t.

$$3x_1 + \quad x_2 + 2x_3 \qquad\qquad = 30$$
$$2x_1 + \quad x_2 + 3x_3 + \quad x_4 \ge 15$$
$$2x_2 \qquad\qquad + 3x_4 \le 25$$
$$x_1,\, x_2,\, x_3,\, x_4 \ge 0$$

a. Find the optimal solution.

b. Calculate the range of optimality or range of insignificance (whichever is appropriate) for c_3.

c. What would be the effect of a four unit decrease in c_3 (from 5 to 1) on the optimal solution and the value of that solution?

d. Calculate the range of optimality or range of insignificance (whichever is appropriate) for c_2.

e. What would be the effect of a three unit increase in c_2 (from 1 to 4) on the optimal solution and the value of that solution?

14. Consider the following linear programming problem.

$$\max 10x_1 + 9x_2 + 4x_3 + 6x_4$$

s.t.

$$3x_1 + 2x_2 + 4x_3 + 2x_4 \leq 70$$
$$5x_1 + 5x_2 + x_3 + 3x_4 \leq 60$$
$$5x_1 + 6x_2 + 3x_3 + x_4 \leq 25$$
$$x_1, x_2, x_3, x_4 \geq 0.$$

a. Find the optimal solution.

b. What are the values of the dual variables?

c. Solve the Dual of the above problem.

d. Verify that the values of the optimal primal variables are given by the negative of the $(c_j - z_j)$ associated with the dual surplus variables.

e. Verify that complementary slackness holds for the optimal Primal and Dual solutions.

15. Consider the Par, Inc. problem which is formulated below.

$$\max 10x_1 + 9x_2$$

s.t.

$\frac{7}{10}x_1 + \phantom{\frac{5}{6}}x_2 \leq 630$	Cutting And Dyeing
$\frac{1}{2}x_1 + \frac{5}{6}x_2 \leq 600$	Sewing
$x_1 + \frac{2}{3}x_2 \leq 708$	Finishing
$\frac{1}{10}x_1 + \frac{1}{4}x_2 \leq 135$	Inspection And Packaging
$x_1, x_2 \geq 0$	

The final tableau is

BASIS	c_j	x_1	x_2	s_1	s_2	s_3	s_4	
		10	9	0	0	0	0	
x_2	9	0	1	30/16	0	−210/160	0	252
s_2	0	0	0	−15/16	1	25/160	0	120
x_1	10	1	0	−20/16	0	300/160	0	540
s_4	0	0	0	−11/32	0	45/320	1	18
z_j		10	9	70/16	0	111/16	0	7668
$c_j - z_j$		0	0	−70/16	0	−111/16	0	

a. Calculate the range of optimality for the profit contribution of the Deluxe bag (c_2).

b. If the profit per Deluxe bag drops to $7 per unit, how will the optimal solution be affected?

c. What unit profit would the Deluxe bag have to have before Par would consider changing its current production plan?

d. If the profit of the Deluxe bags can be increased to $15 per unit, what is the optimal production plan? State what you think will happen before you compute the new optimal solution.

16. In section 4.6, we found the range of feasibility for the right-hand side of the Par, Inc. cutting and dyeing constraint to be $495.6 \le b_1' \le 682\frac{8}{11}$. Using the graphical procedure

a. Show what happens as the number of available hours of cutting and dyeing time is changed from the original 630 hours. Specifically, show the feasible region with 650 hours and then 550 hours.

b. What are the optimal solutions and profits for the two cases in part a.? Can you use the dual variable $u_1 = \frac{79}{16}$ to compute the profits associated with the two cases?

c. Describe graphically and verbally what happens when the cutting and dyeing time extends beyond its upper or lower limit.

17. For the Par, Inc. problem:

    a. Calculate the range of feasibility for b_2 (sewing capacity).

    b. Calculate the range of feasibility for b_3 (finishing capacity).

    c. Calculate the range of feasibility for b_4 (inspection and packaging).

    d. Which of these three departments are you interested in scheduling for overtime? Explain.

    e. Management can schedule overtime in the finishing department for a premium of $4.50 per hour over the current rate. Does it make sense to schedule this overtime? If so, how much overtime can be used? What is the new optimal solution and profit after this overtime is used?

    f. What happens when the range of feasibility for the finishing capacity is exceeded?

18. a. Calculate the final Simplex tableau for the Par, Inc. problem after increasing b_1 from 630 to $682\frac{4}{11}$.

    b. Would the current basis be optimal if b_1 was increased further? If not what would be the new optimal basis?

19. Also for the Par, Inc. problem

    a. How much would profit increase if an additional 30 hours became available in the cutting and dyeing department (i.e., b_1 was increased from 630 to 660)?

    b. How much would profit decrease if 40 man-hours were removed from the sewing department and used elsewhere (say in making saddles)?

    c. How much would profit decrease if because of an employee accident there was only 570 hours instead of 630 available in the cutting and dyeing department?

20. Below are additional conditions encountered by Par, Inc.

    a. Suppose because of some new machinery Par, Inc. was able to make a small reduction in the amount of time it took to do the cutting and dyeing (constraint 1) for a standard bag. What effect would this have on the objective function?

b. Suppose that by buying a new sewing machine management believes the sewing time for Standard bags can be reduced from ½ hour to ⅓ hour. Do you think this machine would be a good investment? Why?

21. Consider the product mix model presented in problem 1. The final Simplex tableau is shown below.

		x_1	x_2	x_3	x_4	s_1	s_2	s_3	
BASIS	c_j	4	6	3	1	0	0	0	
x_3	3	3/60	0	1	1/2	3/10	0	−2/10	125
s_2	0	39/12	0	0	−1/2	−5/10	1	−1	425
x_2	6	39/60	1	0	1/2	−1/10	0	12/30	25
	z_j	81/20	6	3	9/2	3/10	0	54/30	525
	$c_j - z_j$	−1/20	0	0	−7/2	−3/10	0	−54/30	

a. Calculate the range of feasibility for b_1.

b. Calculate the range of feasibility for b_2.

c. Calculate the range of feasibility for b_3.

d. Suppose extra machine hours can be obtained at the following cost:

Machine	Added Cost/hr.
A	$1.00
B	$2.00
C	$1.50

which resource, if any, would you want, how much would you want, and what is the new solution.

22. Consider the following final Simplex tableau.

BASIS	c_j	x_1 3	x_2 1	x_3 5	x_4 3	a_1 $-M$	s_1 0	a_2 $-M$	s_2 0	
s_1	0	5/2	7/6	0	0	3/2	1	-1	1/3	115/3
x_3	5	3/2	1/2	1	0	1/2	0	0	0	15
x_4	3	0	2/3	0	1	0	0	0	1/3	25/3
	z_j	15/2	9/2	5	3	5/2	0	0	1	100
	c_j-z_j	$-9/2$	$-7/2$	0	0	$-5/2-M$	0	$-M$	-1	

The original formulation for this problem is shown in problem 4.

a. Calculate the range of feasibility for b_1.

b. Calculate the range of feasibility for b_2.

c. Calculate the range of feasibility for b_3.

5

Applications of Linear Programming

Our study thus far has been directed primarily toward obtaining an understanding of linear programming methodology. This background is essential for knowing when linear programming is an appropriate problem-solving tool and for interpreting the results of a linear programming solution to a problem. However, the benefits of this study will only be realized by learning how this methodology can be used to solve real-world decision-making problems. The purpose of this chapter is to show you how selected real-world decision-making problems can be formulated and solved using linear programming.

There are two ways in which one may develop skills in model building. (In this book model building should be taken to mean formulating a linear program that is a "model" of the real-world decision-making problem for which a solution is desired.) The first way is by on-the-job experience. This is essentially a trial-and-error approach and obviously could not be attempted in a textbook. The second way in which one may develop these skills is by studying the way in which others have developed successful models. Thus, in this chapter we attempt to develop the reader's skills along these lines by presenting several reasonably detailed examples of successful linear programming applications.

In practice, linear programming has proven to be one of the most successful quantitative aids for managerial decision making. Numerous applications have been reported in the chemical, airline, steel, paper, petroleum, and other industries. The specific problems studied have included production scheduling, capital budgeting, plant location, transportation, media selection, and many others.

As the variety of the applications mentioned would suggest, linear programming is a flexible problem-solving tool with applications in many disciplines. In this chapter we present introductory applications from the areas of finance, management, and marketing, as well as the standard linear programming applications in transportation, assignment, blending, and diet problems. In addition, an application involving environmental protection is presented.

An understanding of the material presented in this chapter should give the reader an appreciation of the broad range of practical linear programming applications and provide a basis for the reader to further develop his modeling skills by creating similar and possibly new linear programming applications in his own field of interest.

5.1   Financial Applications

Portfolio Selection

Portfolio selection problems are financial management situations in which a manager must select specific investments—stocks, bonds, etc.—from a variety of investment alternatives. This type of problem is frequently encountered by managers of mutual funds, credit unions, insurance companies, banks, etc. The objective function for these problems is usually maximization of expected return or minimization of risk. The constraints usually take the form of restrictions on the type of permissible investments, state laws, company policy, maximum permissible risk, etc.

Problems of this type have been formulated and solved using a variety of mathematical programming techniques. However, if in a particular portfolio selection problem it is possible to formulate a linear objective function and linear constraints (i. e., if the assumptions of proportionality and additivity discussed in Chapter 1 are satisfied), then linear programming can be used to solve the problem.   In this section we show how a simplified portfolio selection problem can be formulated and solved as a linear program.[1]

Consider the case of Welte Mutual Funds, Inc. located in New York City.   Welte has just obtained $100,000 by converting industrial bonds to cash and is now looking for other investment opportunities for these funds.   Considering Welte's current investments, the firm's top financial analyst recommends that all new investments should be made in the oil industry, steel industry, or government bonds.   Specifically, the analyst has identified five investment opportunities and projected their annual rates of return.   The investments and rates of return are shown in Table 5–A.

Investment	Projected Rate of Return
Atlantic Oil	7.3%
Pacific Oil	10.3%
Midwest Steel	6.4%
Huber Steel	7.5%
Government Bonds	4.5%

Table 5–A.   Investment Opportunities for Welte Mutual Funds

Management of Welte has imposed the following investment guidelines:

1.   neither industry should receive more than 50% of the total new investment;

2.   government bonds should be at least 25% of the steel industry investments; and

---

[1] For a discussion of some other approaches to Portfolio selection, see: Markowitz, H., *Portfolio Selection* (Cowles Foundation Monograph No. 16). New York: Wiley, 1959.

3. the investment in Pacific Oil, the high return but high risk investment, cannot be more than 60% of the total oil industry investment.

What portfolio recommendations—investments and amounts—should be made for the available $100,000? Given the objective of maximizing projected return subject to the budgetary and managerially imposed constraints, we can answer this question by formulating a linear programming model of the problem. The solution to this linear programming model will then provide investment recommendations for the management of Welte Mutual Funds.

Let

$x_1 = $ dollars invested in Atlantic Oil,

$x_2 = $ dollars invested in Pacific Oil,

$x_3 = $ dollars invested in Midwest Steel,

$x_4 = $ dollars invested in Huber Steel, and

$x_5 = $ dollars invested in Government Bonds.

The complete linear programming model is as follows:

max $.073x_1 + .103x_2 + .064x_3 + .075x_4 + .045x_5$

s. t.

$x_1 +$	$x_2 +$	$x_3 +$	$x_4 +$	$x_5 =$	100,000	Available Funds
$x_1 +$	x_2				$\leq$ 50,000	Oil Industry Maximum
		$x_3 +$	x_4		$\leq$ 50,000	Steel Industry Maximum
		$- .25x_3 -$	$.25x_4 +$	$x_5 =$	0	Government Bonds Minimum
$-.6x_1 +$	$.4x_2$				$\leq$ 0	Pacific Oil Restriction

$x_1, x_2, x_3, x_4, x_5 \geq 0$.

The solution to this linear programming model is shown in Table 5–B.

Investment	Amount	Expected Annual Return
Atlantic Oil	$ 20,000	$ 1,460
Pacific Oil	30,000	3,090
Huber Steel	40,000	3,000
Government Bonds	10,000	450
	$100,000	$ 8,000

Expected annual return—$8,000 (8%).

Table 5–B. Optimal Portfolio Selection
for Welte Mutual Funds

We note that the optimal solution indicates that the portfolio should be diversified among all of the investment opportunities except Midwest Steel. The projected expected annual return for this portfolio is 8%.

One shortcoming of the linear programming approach to the portfolio selection problem is that we may not be able to invest the exact amount specified in each of the securities. For example, if Atlantic Oil sold for $75 a share we would have to purchase exactly 266 and ⅔ shares in order to spend exactly the recommended $20,000. The approach usually taken to avoid this difficulty is to purchase the largest possible whole number of shares with the amount of funds recommended (e. g., 266 shares for Atlantic Oil). Hence, we guarantee that our budget constraint will not be violated. This, of course, introduces the possibility that our solution will no longer be optimal, but the danger is slight if large numbers of securities are involved.

Financial Mix Strategy

Financial mix strategies involve the selection of means for financing company projects, inventories, production operations and various other activities. In this section we illustrate how linear programming can be used to solve problems of this type by formulating and solving a problem involving the financing of production operations. In this particular application a financial decision must be made with regard to how much production is to be supported by internally generated funds and how much is to be supported by external funds.

The Jefferson Adding Machine Company will begin production of two new models of electronic adding machines during the next three months. Since this new line requires an expansion of the current production operation, the Company will need operating funds to cover material, labor, and selling expenses during this initial production period. Revenue from this initial production run will not be available until after the end of the period; thus the company must arrange financing for these operating expenses before production can begin.

Jefferson has $3,000 in internal funds available to cover expenses of this operation; if additional funds are needed, they will have to be generated externally. A local bank has offered a line of short-term credit in an amount not to exceed $10,000. The interest rate over the life of the loan will be 12% per year on the average amount borrowed. One stipulation set by the bank requires that the total of the company cash allocated to this operation plus the accounts receivable for this product line must be at least twice as great as the outstanding loan plus interest at the end of the initial production period.

In addition to the financial restrictions placed on this operation, man-hour capacity is also a factor for Jefferson to consider. Specifically, only 2500 hours of assembly time and 150 hours of packaging and shipping time are available for the new product line during the initial three-month production period. Relevant cost, price, and production time requirements for the two models are shown in Table 5–C.

Model	Unit Cost (Mat'ls & other variable exp.)	Selling Price	Profit Margin	Man-Hours Required Assembly Dept.	Packaging & Shipping
Y	$ 50	$ 58	$ 8	12	1
Z	$100	$120	$20	25	2

Table 5–C. Cost, Price, and Manpower Data for the Jefferson Adding Machine Company

Additional restrictions have been imposed by company management in order to guarantee that the market reaction to both products can be tested. That is, at least 50 units of model Y and at least 25 units of model Z must be produced in this first production period.

Since the cost of units produced on borrowed funds will in effect experience an interest charge, the profit margins for the units of models Y and Z produced on borrowed funds will be reduced. Hence, we adopt the following notation for the decision variables in our problem:

$x_1 =$ units of model Y produced with company funds,

$x_2 =$ units of model Y produced with borrowed funds,

$x_3 =$ units of model Z produced with company funds, and

$x_4 =$ units of model Z produced with borrowed funds.

How much will the profit margin be reduced for units produced on borrowed funds? To answer this question, one must know for how long the loan will be outstanding. We assume that all units of each model are sold as they are produced to independent distributors, and that the average rate of turnover of accounts receivable is three months. Since company management has specified that the loan is to be repaid by funds generated by the units produced on borrowed funds, the funds borrowed to produce one unit of model Y or Z will be repaid approximately three months later. Hence, the profit margin for each unit of model Y produced on borrowed funds is reduced from $8.00 to $6.50 = $8.00 − ($50x.12x¼ yr.), and the profit margin for each unit of model Z produced on borrowed funds is reduced from $20.00 to $17.00 = $20.00 − ($100x.12x¼ yr.). With this information we can now formulate the objective function for Jefferson's financial mix problem.

$$\max 8x_1 + 6.5x_2 + 20x_3 + 17x_4$$

We can also specify the following constraints for the model.

$12x_1 + 12x_2 + 25x_3 + 25x_4 \leq 2500$				Assembly Dept.
$x_1 + x_2 + 2x_3 + 2x_4 \leq 150$				Packaging & Shipping Dept.
$50x_1 + 100x_3 \leq 3000$				Internal Funds
$50x_2 + 100x_4 \leq 10000$				External Funds
$x_1 + x_2 \geq 50$				Model Y Requirement
$x_3 + x_4 \geq 25$				Model Z Requirement

In addition, the following constraint must be included to satisfy the bank loan requirement.

$$\text{Cash} + \text{Accounts Receivable} \geq 2 \,(\text{Loan} + \text{Interest})$$

This restriction must be satisfied at the end of the period. Recalling that accounts receivable are outstanding for an average of three months the following relationships can be used to derive a mathematical expression for the above inequality at the end of the period.

$$\text{Cash} = 3000 - 50x_1 - 100x_3$$
$$\text{Accounts Receivable} = 58x_1 + 58x_2 + 120x_3 + 120x_4$$
$$\text{Loan} = 50x_2 + 100x_4$$
$$\text{Interest} = (.12\text{x}\tfrac{1}{4} \text{ yr.}) \,(50x_2 + 100x_4) = 1.5x_2 + 3x_4$$

Therefore, the constraint resulting from the bank restriction can be written as:

$$3000 - 50x_1 - 100x_3 + 58x_1 + 58x_2 + 120x_3 + 120x_4 \geq 2\,(51.5x_2 + 103x_4),$$

or

$$-8x_1 + 45x_2 - 20x_3 + 86x_4 \leq 3000.$$

The complete linear programming model for our problem can now be stated.

$$\max \quad 8x_1 + 6.5x_2 + 20x_3 + 17x_4$$

s.t.

$$
\begin{aligned}
12x_1 + 12x_2 + 25x_3 + 25x_4 &\leq 2500 \\
x_1 + x_2 + 2x_3 + 2x_4 &\leq 150 \\
50x_1 + 100x_3 &\leq 3000 \\
50x_2 + 100x_4 &\leq 10000 \\
x_1 + x_2 &\geq 50 \\
x_3 + x_4 &\geq 25 \\
-8x_1 + 45x_2 - 20x_3 + 86x_4 &\leq 3000 \\
x_1, x_2, x_3, x_4 &\geq 0
\end{aligned}
$$

The solution to this four-variable, seven-constraint financial mix problem is shown in Table 5–D.

	Units	Expected Profit
Model Y		
Borrowed Funds (x_2)	50	$ 325
Model Z		
Company Funds (x_3)	30	$ 600
Borrowed Funds (x_4)	15.7	$ 267
	Total	$1192

Table 5–D. Optimal Financial Mix for the Production of Jefferson Adding Machines

The optimal financial mix requires the company to use all of its internal funds ($3000), but only slightly over $4000 of the available $10,000 line of credit. Since the profit margins are positive for units produced on borrowed funds, why doesn't the optimal linear programming solution call for the company to borrow more funds, produce more units, and thus make a larger profit? The obvious reason is that to do so would violate one or more of the constraints in the problem. Upon closer examination, we see that there are actually two binding constraints at the optimal solution; i. e., both the internal funds constraint and the bank requirement constraint (ignoring the small rounding error of 0.2) are binding.

Thus, from this linear programming analysis, company management has learned that it can improve the profit picture if additional internal funds can be made available for the project and/or other lending sources with less stringent requirements can be found.

As a final note concerning this financial mix problem, you may recall we stated in Chapter 4 that the Dual of a linear programming model can provide valuable information on the sensitivity of the objective function to changes in the right-hand sides. Remember that before converting to the Dual we must write our original problem in its equivalent Primal form. As shown in Chapter 4, this requires changing the two greater-than-or-equal-to constraints in our formulation. Specifically we multiply through by (-1) to obtain

$$- x_1 - x_2 \qquad \leq - 50$$
$$- x_3 - x_4 \leq - 25.$$

The complete Dual formulation of our financial mix problem can then be written as follows:

$$\min 2500u_1 + 150u_2 + 3000u_3 + 10000u_4 - 50u_5 - 25u_6 + 3000u_7$$

s.t.

$$
\begin{array}{llllllll}
12u_1 + & u_2 + & 50u_3 & & - u_5 & & & 8u_7 \geq 8 \\
12u_1 + & u_2 & & + & 50u_4 & - u_5 & + & 45u_7 \geq 6.5 \\
25u_1 + & 2u_2 + & 100u_3 & & & - u_6 - & & 20u_7 \geq 20 \\
25u_1 + & 2u_2 & & + & 100u_4 & - u_6 + & & 86u_7 \geq 17
\end{array}
$$

$$u_1, u_2, u_3, u_4, u_5, u_6, u_7 \geq 0.$$

The optimal solution to the Dual is given by:

$$u_1 = 0 \qquad\qquad u_5 = 2.40$$
$$u_2 = 0 \qquad\qquad u_6 = 0$$
$$u_3 = .24 \qquad\qquad u_7 = .20$$
$$u_4 = 0$$

From the dual solution we see that $u_3 = .24$. Recalling that value of a dual variable tells us how much the objective function value changes for a one unit change in the right-hand side of the corresponding constraint, we conclude that every additional dollar of internally generated funds will increase profits by $0.24. Of course, as is always the case, this rate of increase is only applicable as long as the current basis remains optimal. We also note that $u_7 = .20$. Thus, every unit change in the right-hand side of the bank loan requirement constraint will increase profits by $0.20. But, we saw that the right-hand side of the bank loan requirement constraint was $3000, the amount of internal funds available. Thus, with respect to the bank loan constraint, we see that every additional dollar of internal funds works to improve profit even more than the $0.24 indicated when analyzing just the internal funds constraint. In fact, every additional dollar of internally generated funds will increase profits by $0.44 = .24 + .20 = u_3 + u_7$. Management analysis of this 44% marginal return on internally generated funds might lead the company to reallocate internal funds from some less profitable venture in order to finance production operations for this new product line.

Observe that dual variable 5 indicates that an increase in the right-hand side of the Primal form of constraint 5 by one unit will lead to a $2.40 increase in profit as long as the same basis remains optimal. (Unless the Primal solution is degenerate it is safe to assume that a small change in a right-hand side will not affect the optimal basis.) The Primal form of constraint 5 was given by

$$- x_1 - x_2 \quad \leq - 50.$$

Hence, we note that a one unit increase in this right-hand side corresponds to requiring that only 49 units of model Y must be produced. Given this information management might elect to relax this requirement. Thus, we see that the solution to the Dual problem has provided the company's management with additional information as to how the profit picture might be improved.

5.2   Marketing Applications

Media Selection

Media selection applications of linear programming are aimed at helping marketing managers allocate a fixed advertising budget across various media. Potential advertising media include newspapers, magazines, radio commercials, television commercials, direct mailings and others. In most of these applications the objective is taken to be the maximization of audience exposure. Restrictions on the allowable allocation usually arise through considerations such as company policy, contract requirements, availability of media, etc. In the application which follows we illustrate how a simple media selection problem might be formulated and solved using a linear programming model.

Consider the case of the Relax-and-Enjoy Lake Development Corporation. Relax is developing a lake-side community at a privately owned lake and is in the business of selling property for vacation and/or retreat cottages. The primary market for these lake-side lots includes all middle and upper income families within approximately one-hundred miles of the development. Relax-and-Enjoy has employed the advertising firm of Boone, Phillips and Jackson to design the promotional campaign for the project.

After considering possible advertising media and the market to be covered, Boone has made the preliminary recommendation to restrict the first month's advertising to five sources. At the end of this month, Boone will then reevaluate its strategy based upon this month's results. Boone has collected data on the number of potential purchase families reached, the cost per advertisement, the maximum number of times each media is available, and the expected exposure for each of the five media. The expected exposure is measured in terms of an exposure unit, a management judgment measure of the relative value of one advertisement in each of the media. These measures, based on Boone's experience in the advertising business, take into account such factors as audience profile (e.g., age, income, and education of the audience reached), image presented, and quality of the advertisement. The information collected to date is presented in Table 5-E.

Advertising Media	Number of Potential Purchase Families Reached	Cost per Advertisement	Maximum * Times Available per month	Expected Exposure Units
1) Daytime TV (1 min) Station WKLA	1,000	$1,500	15	65
2) Evening TV (30 secs) Station WKLA	2,000	$3,000	10	90
3) Daily Newspaper (Full Page) The Morning Journal	1,500	$400	25	40
4) Sunday Newspaper Magazine (½ page-color) The Sunday Press	2,500	$1,000	4	60
5) Radio (30 secs 8:00 a. m. or 5:00 p. m. News) Station KNOP	300	$100	30	20

* The maximum number of times the media is available is either the maximum number of times the advertising media occurs (i. e., 4 Sundays for media 4) or the maximum number of times Boone will allow the media to be used.

Table 5-E.   Advertising Media Alternatives for the Relax-and-Enjoy, Lake Development Corporation

Relax-and-Enjoy has provided Boone with an advertising budget of $30,000 for the first month's campaign. In addition, Relax-and-Enjoy has imposed the following restrictions on how Boone may allocate these funds. At least ten television commercials must be used and at least 50,000 potential purchasers must be reached during the month. In addition no more than $18,000 may be spent on television advertisements. What advertising media selection plan should the advertising firm recommend?

The first step in formulating a linear programming model of this problem is the introduction of suitable notation. We let

$x_1 = $ number of times daytime TV is used,

$x_2 = $ number of times evening TV is used,

$x_3 = $ number of times daily newspaper is used,

$x_4 = $ number of times Sunday newspaper is used, and

$x_5 = $ number of times radio is used.

With the overall goal of maximizing expected exposure, the objective function becomes

$$\max 65x_1 + 90x_2 + 40x_3 + 60x_4 + 20x_5.$$

The constraints for our model can now be formulated from the information given.

$$
\begin{array}{ll}
x_1 \qquad\qquad\qquad\qquad\qquad\qquad\qquad \le 15 & \left.\begin{array}{l}\\\\\\\\\\\end{array}\right\} \\
\qquad x_2 \qquad\qquad\qquad\qquad\qquad\qquad \le 10 & \text{Availability} \\
\qquad\qquad x_3 \qquad\qquad\qquad\qquad\qquad \le 25 & \text{of Media} \\
\qquad\qquad\qquad x_4 \qquad\qquad\qquad \le 4 & \\
\qquad\qquad\qquad\qquad x_5 \le 30 & \\
1500x_1 + 3000x_2 + 400x_3 + 1000x_4 + 100x_5 \le 30000 & \text{Budget} \\
x_1 + \qquad x_2 \qquad\qquad\qquad\qquad \ge 10 & \left.\begin{array}{l}\text{Television}\\\text{Restrictions}\end{array}\right. \\
1500x_1 + 3000x_2 \qquad\qquad\qquad \le 18000 & \\
1000x_1 + 2000x_2 + 1500x_3 + 2500x_4 + 300x_5 \ge 50000 & \begin{array}{l}\text{Audience}\\\text{Coverage}\end{array}
\end{array}
$$

$$x_1, x_2, x_3, x_4, x_5 \ge 0$$

The solution to this five-variable, nine-constraint linear programming model is presented in Table 5–F.

Media	Frequency	Budget
Daytime TV	10	$15,000
Daily Newspaper	25	10,000
Sunday Newspaper	2	2,000
Radio	30	3,000
		$30,000

Total Audience Contacted—61,500

Expected Exposure—2,370

Table 5–F. Advertising Plan for Relax-and-Enjoy,
Lake Development Corporation

We should point out that the above media selection model, probably more than most other linear programming models, requires crucial subjective evaluations as input. The most critical of these inputs is the expected exposure rating measure. While marketing managers may have substantial data concerning expected advertising exposure, the final coefficient that includes image and quality considerations is primarily based on managerial judgment. However, judgment input is a very acceptable way of obtaining necessary data for a linear programming model.

Another shortcoming of this model is that even if the expected exposure measure was not subject to error, there is no guarantee that maximization of total expected exposure will lead to a maximization of profit, or sales (a common surrogate for profit). However, this is not a shortcoming of linear programming, rather it is a shortcoming of the use of exposure as a criterion. Certainly if we were able to measure directly the effect of an advertisement on profit we would use total profit as our objective to be maximized.

In addition, you should be aware that the media selection model as formulated in this section does not include considerations such as the following:

1. reduced exposure value for repeat media usage,

2. cost discounts for repeat media usage,

3. audience overlap by different media,

4. timing recommendation for the advertisement.

A more complex formulation—more variables and constraints —can often be used to overcome some of these limitations, but it will not always be possible to overcome all of them with a linear programming model. However, even in these cases, a linear programming model can often be used to arrive at a rough approximation to the best decision. Management evaluation combined with the linear programming solution should then make possible the selection of an overall effective advertising strategy.

Marketing Strategy

One particular marketing strategy decision involves the optimal allocation of salesforce and advertising effort. As we discussed in the previous section on media selection problems, one would like to make this decision in such a fashion as to maximize profit or sales. Unfortunately, one seldom has enough information to specify the relationship between the allocation of salesforce and advertising effort and the ultimate criterion of profit or sales. We illustrate in this section a case where the company has been able to specify this relationship; thus the marketing strategy decision can be made with the objective of maximizing profit.

Electronic Communications, Inc. manufactures portable radio systems that can be used for two-way communications. The company's new product, which has a range of up to 25 miles, is particularly suitable for use in a variety of applications; e. g., mobile unit-home office systems, marina sales and service systems, etc. In these applications the two-way communication system enables an office to easily contact field salesmen, repairmen, etc. The primary distribution channel for the product will be through industrial communications equipment distributors; however, the firm is also considering distribution through a national chain of discount stores and a marine equipment distributor. These latter two distribution channels have the advantage of allowing the product to reach individuals interested in radio-oriented hobbies and individuals desiring boat communication systems.

Because of differing distribution and promotional costs, the profitability of the product varies with the distribution alternative selected. In addition, the company's estimate of the advertising cost and salesman time per unit sold will vary with the different distribution channels. Since the company only produces these units on order, the number of units produced and number of units sold are the same.

Table 5–G summarizes the data prepared by Electronic Communications with respect to profit, expected advertising effort per unit sold, and estimated salesforce effort per unit sold. The advertising and salesforce estimates are based upon past experience with similar radio equipment.

Distributor	Profit per Unit Sold	Estimated Average Advertising Effort per Unit Sold	Estimated Salesforce Effort per Unit Sold (Hours)
Industrial	$90	$10	2.5
Discount Stores	70	18	3.0
Marine	84	8	3.0

Table 5–G. Profit, Cost, and Time Data
for Electronic Communications, Inc.

Company management has specified that at least 100 units must be distributed through the discount stores during the next three months. The firm has set the advertising budget at $5000 and stated that a maximum of 1200 man-hours of salesforce time will be available during the current planning period. In addition, production capacity is 600 units.

The company is now faced with the task of establishing a profitable marketing strategy. Specifically, decisions need to be made on the following:

1. how many units should be produced and how should they be allocated to the three market segments;

2. how much advertising should be devoted to the three market segments;

3. how should the salesforce effort be allocated among the three market segments.

Proceeding to a linear programming formulation of this problem we introduce the following notation:

x_1 = units produced for the industrial market,

x_2 = units produced for the discount store market, and

x_3 = units produced for the marine market.

In terms of this notation, the objective function and constraints can be written as follows:

$$\max 90x_1 + 70x_2 + 84x_3$$

s.t.

$$10x_1 + 18x_2 + 8x_3 \leq 5000 \quad \text{Advertising Budget}$$
$$2.5x_1 + 3x_2 + 3x_3 \leq 1200 \quad \text{Salesforce Availability}$$
$$x_1 + x_2 + x_3 \leq 600 \quad \text{Production Capacity}$$
$$x_2 \geq 100 \quad \text{Minimum Discount Store Volume}$$
$$x_1, x_2, x_3 \geq 0.$$

The solution to this linear programming model is given in Table 5–H.

Market Segment	Volume	Advertising Allocation	Salesforce Allocation
Industrial	240	$ 2,400	600 hrs.
Discount Store	100	1,800	300 hrs.
Marine	100	800	300 hrs.
Total	440	$ 5,000	1200 hrs.

Profit Projection—$37,000

Figure 5–H.   Profit Maximizing Marketing Strategy for Electronic Communications, Inc.

Consideration of the dual variables and application of sensitivity analysis techniques may provide the marketing manager with some additional valuable information. Specifically, the dual variables for the first three resource constraints are as follows: $u_1 = 6$, $u_2 = 12$, and $u_3 = 0$. Recall from Chapter 4 that a dual variable of zero indicated that an increase in the value of the right-hand side for the corresponding Primal constraint would have no effect on profit. Since $u_3 = 0$, we can conclude in this case that the production capacity (constraint # 3) is not restricting our profit. In fact the slack variable associated with this constraint shows the excess production capacity to be 160 units.

Recall that the non-zero dual variables mean that the corresponding primal constraints hold as equalities and are thus binding. Since $u_1 = 6$, and $u_2 = 12$, we know that we are using the maximum available advertising and salesforce resources. The results in Table 5–H confirm this analysis. The values of these dual variables give the marginal value of additional advertising budget and salesforce effort. Specifically, an additional advertising dollar has a potential of increasing the profit by six dollars ($u_1 = 6$) while an additional man-hour of salesforce effort has a potential value of twelve dollars ($u_2 = 12$). Thus, the manager should consider the possibility of obtaining these additional resources as long as the cost of the addition is less than the potential benefits. Recall however that we saw in Chapter 4 that we cannot expect additional resources to increase profit without limit. In this case we might consider increasing the advertising budget and using part-time sales assistance; however, as we continue to increase these resources, sales will increase and production capacity will become binding causing any additional advertising and sales efforts to be of no value.

5.3 Management Applications

Production Scheduling

One of the richest areas of linear programming applications is production scheduling. The solution to a production scheduling problem enables the manager to establish an efficient low-cost production schedule for one or more products over several time periods; e. g., weeks, months, etc. Essentially, a production scheduling problem can be viewed as a product mix problem for each of several periods in the future. The manager must determine the production levels that will allow the company to meet product demand requirements given limitations on production capacity, manpower capacity, and storage space. At the same time, it is desired to minimize the total cost of carrying out this task.

One major reason for the wide-spread application of linear programming to production scheduling problems is that these problems are of a recurring nature. A production schedule must be established for the current month, then again for the next month, the month after that, and so on. When the production manager looks at the problem each month he will find that while demands for his products have changed, production times, production capacities, storage space limitations, etc., are roughly the same; thus, he is basically resolving the same problem he handled in previous months. Hence, a general linear programming model of the production scheduling procedure may be frequently applied. Once the model has been formulated, the manager can simply supply the data—demands, capacities, etc.—for the given production period, and the linear programming model can then be used to develop the production schedule. Thus one linear programming formulation may have many repeat applications.

In Chapter 6, we present in detail an actual large-scale production scheduling application that involves establishing a monthly production plan for 500 products. The basic problem formulation concepts will be identical to the ones used in the rather small production scheduling problem presented in this section. Thus, a thorough understanding of the following application will be a valuable aid in the study of the large-scale application presented in Chapter 6.

Let us consider the case of the Bollinger Electronics Company which produces two different electronic components for a major airplane engine manufacturer. The airplane engine manufacturer notifies the Bollinger sales office each quarter as to what the monthly requirements for components will be during each of the next three months. The monthly demands for the components may vary considerably depending upon the type of engines the airplane engine manufacturer is producing. The order shown in Table 5–I has just been received for the next three-month period.

Order:

General Engine Manufacturing Company

	Month		
	April	May	June
Component 322A	1,000	3,000	5,000
Component 802B	1,000	500	3,000

Table 5–I. Three-Month Demand Schedule for Bollinger Electronics Company

After the order is processed, a demand statement is sent to the production control department. The production control department must then develop a three-month production plan for the components. Knowing the preference of the production department manager for constant demand levels (balanced workload, constant machine and manpower utilization, etc.), the production scheduler might consider the alternative of producing at a constant rate for all three months. This would set monthly production quotas at 3,000 units per month for component 322A and 1,500 units per month for component 802B. Why not adopt this schedule?

While this schedule would obviously be quite appealing to the production department, it may be undesirable from a total cost point of view. In particular this schedule ignores inventory costs. Consider the projected inventory levels that would result from this schedule calling for constant production (Figure 5–A).

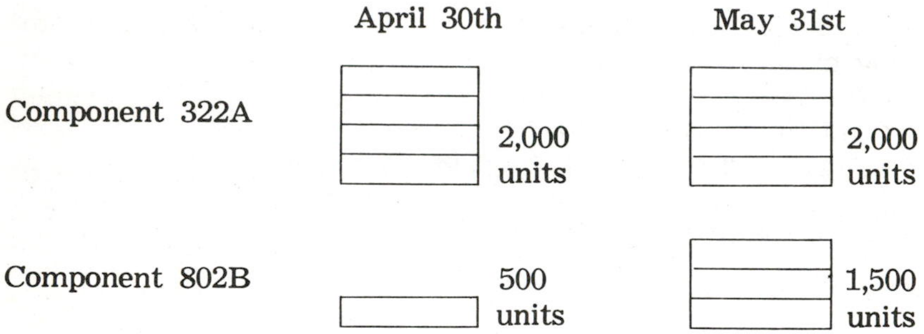

	April 30th	May 31st
Component 322A	2,000 units	2,000 units
Component 802B	500 units	1,500 units

(We are assuming no beginning inventory for either component at the start of the three month period—April 1.)

Figure 5–A.   Projected Inventory Levels Under a Constant Rate Production Schedule

Thus, we see that this production schedule would lead to high inventory levels. When we consider the cost of tied-up idle capital and storage space, a schedule that provides lower inventory levels might be economically more desirable.

At the other extreme of the constant rate production schedule is the produce-to-meet-demand approach. While this schedule eliminates the inventory holding cost problem, the wide monthly production level fluctuations may cause some serious production problems and costs. For example, production capacity would have to be available to meet the total 8,000 unit peak demand in June.   Unless other components could be scheduled on the same production equipment in April and May, there would be significant unused capacity and thus low machine utilization in those months.   In addition, the large production variations will require substantial manpower adjustments; employee turnover or training problems may be encountered.   Thus it appears that the best production schedule will be one that compromises between the constant rate-high inventory and the variable rate-low utilization extremes.

The production scheduler will therefore want to identify and consider the following costs:

1.   production costs,
2.   storage costs, and
3.   change in production level costs.

In the remainder of this section we show how a linear programming model of the production process for Bollinger Electronics can be formulated to account for these costs in such a fashion that the total system cost is minimized.

In order to develop our model we introduce a double subscript notation for the decision variables in the problem.   We let the first subscript indicate the product number and the second subscript indicate the month.   Thus in general we have

$x_{im} =$ production volume in units for product i in month m. Here i = 1, 2, and m = 1, 2, 3; i = 1 refers to component 322A, i = 2 refers to component 802B.

The purpose of the double subscript is to provide a more descriptive notation.   We could simply use x_6 to represent the number of units of product 2 produced in month 3, but we believe (as do many others) that x_{23} is more descriptive in that we know directly the product and month the variable represents.

If component 322A costs $20.00 per unit produced and component 802B costs $10.00 per unit produced, the production cost part of the objective function becomes

Production Cost: $20x_{11} + 20x_{12} + 20x_{13} + 10x_{21} + 10x_{22} + 10x_{23}.$

You should note that in this particular problem the production cost per unit is the same each month, and thus we need not include production costs in our objective function; i. e., no matter what production schedule is selected, the total production costs will remain the same. In cases where the cost per unit is expected to change each month, these variable production costs per unit per month must be included in the objective function. For the Bollinger Electronics problem, we have elected to include them. This means that the value of the linear programming objective function will include all of the costs associated with the problem.

To incorporate the inventory cost into our model, we introduce the following double subscripted decision variable that will indicate the number of units of inventory for each product for each month. We let

s_{im} = the inventory level for product i at the end of month m.

Bollinger has determined that, on a monthly basis, inventory holding costs are 1.5% of the value of the product—$.30 per unit for component 322A and $.15 per unit for component 802B. A common assumption made in linear programming approaches to the production scheduling problem is now invoked. That is, monthly ending inventories are an acceptable approximation to the average inventory levels throughout the month. Given this assumption, the inventory holding cost portion of the objective function can be written as follows:

Inventory
Holding Cost: $.30s_{11} + .30s_{12} + .30s_{13} + .15s_{21} + .15s_{22} + .15s_{23}.$

In order to incorporate the costs due to fluctuations in production levels we need to define the following additional decision variables. Let

I_m = increase in the production man-hours during month m, and

D_m = decrease in the production man-hours during month m.

After estimating the effect of employee layoffs, turnovers, reassignment training costs, and other costs associated with fluctuating manpower requirements, Bollinger estimated that the cost associated with an increase in manpower was $10.00 per man-hour, while the cost associated with a decrease was only

$2.50 per man-hour. Thus, the third portion of our objective function can now be written.

Production
Fluctuation
Costs: $10I_1 + 10I_2 + 10I_3 + 2.5D_1 + 2.5D_2 + 2.5D_3$

You should note that Bollinger has elected to measure the cost associated with production fluctuations as a function of the change in man-hours required. In other production scheduling problems such costs might be measured in terms of machine hours or in terms of total units produced.

Combining all three costs our complete objective function becomes:

Objective
Function:
$$20x_{11} + 20x_{12} + 20x_{13} + 10x_{21} + 10x_{22} + 10x_{23} + .30s_{11} + .30s_{12} + .30s_{13} + .15s_{21} + .15s_{22} + .15s_{23} + 10I_1 + 10I_2 + 10I_3 + 2.5D_1 + 2.5D_2 + 2.5D_3.$$

Now let us consider the constraints. First, we must guarantee that our schedule meets customer demand. Since the units shipped can come from the current month's production or from inventory carried over from previous periods, we have the following basic requirement:

$$\begin{pmatrix} \text{Ending} \\ \text{Inventory} \\ \text{from Previous} \\ \text{Month} \end{pmatrix} + \begin{pmatrix} \text{Current} \\ \text{Production} \end{pmatrix} \geq \begin{pmatrix} \text{This Month's} \\ \text{Demand} \end{pmatrix}$$

In fact, the difference in the left-hand side and the right-hand side will be the amount of ending inventory at the end of this month. Thus the demand requirement takes the form

$$\begin{pmatrix} \text{Ending} \\ \text{Inventory} \\ \text{from Previous} \\ \text{Month} \end{pmatrix} + \begin{pmatrix} \text{Current} \\ \text{Production} \end{pmatrix} - \begin{pmatrix} \text{Ending} \\ \text{Inventory} \\ \text{for this} \\ \text{Month} \end{pmatrix} = \begin{pmatrix} \text{This} \\ \text{Month's} \\ \text{Demand} \end{pmatrix}$$

Suppose that the inventories at the beginning of our three-month scheduling period were 500 units for component 322A and 200 units for component 802B. Recalling that the demand for both products in the first month (April) was 1,000 units, the constraints for meeting demand in the first month become

$$500 + x_{11} - s_{11} = 1000$$
$$200 + x_{21} - s_{21} = 1000.$$

Moving the constants to the right-hand side, we have

$$x_{11} - s_{11} = 500$$
$$x_{21} - s_{21} = 800.$$

Similarly, we need demand constraints for both products in the second and third months. These can be written as follows:

Month 2
$$s_{11} + x_{12} - s_{12} = 3000$$
$$s_{21} + x_{22} - s_{22} = 500.$$

Month 3
$$s_{12} + x_{13} - s_{13} = 5000$$
$$s_{22} + x_{23} - s_{23} = 3000.$$

If the company specifies a minimum inventory level at the end of the three-month period of at least 400 units of component 322A and at least 200 units of component 802B, we can add the constraints

$$s_{13} \geq 400$$
$$s_{23} \geq 200.$$

Let us suppose that we have the following additional information available on production, manpower, and storage capacity given in Table 5–J.

	Machine Capacity (Hrs)	Manpower Capacity (Hrs)	Storage Capacity (sq. ft.)
April	400	300	10,000
May	500	300	10,000
June	600	300	10,000

Table 5–J. Machine, Manpower, and Storage Capacities for Bollinger Electronics

Machine, manpower, and storage space requirements are given in Table 5–K.

	Machine Hours/Unit	Man-hours per Unit	Storage sq. ft./unit
Component 322A	.10	.05	2
Component 802B	.08	.07	3

Table 5–K. Machine, Manpower, and Storage Requirements for Components 322A and 802B.

To reflect these limitations the following constraints are necessary.

Machine Capacity

$$.10x_{11} + .08x_{21} \leq 400 \quad \text{Month 1}$$
$$.10x_{12} + .08x_{22} \leq 500 \quad \text{Month 2}$$
$$.10x_{13} + .08x_{23} \leq 600 \quad \text{Month 3}$$

Manpower Capacity

$$.05x_{11} + .07x_{21} \leq 300 \quad \text{Month 1}$$
$$.05x_{12} + .07x_{22} \leq 300 \quad \text{Month 2}$$
$$.05x_{13} + .07x_{23} \leq 300 \quad \text{Month 3}$$

Storage Capacity

$$2s_{11} + 3s_{21} \leq 10000 \quad \text{Month 1}$$
$$2s_{12} + 3s_{22} \leq 10000 \quad \text{Month 2}$$
$$2s_{13} + 3s_{23} \leq 10000 \quad \text{Month 3}$$

One final set of constraints must be added. These are necessary in order to guarantee that I_m and D_m will reflect the increase or decrease in the number of man-hours used for production in month m. Suppose the number of man-hours used for production of the two components in March—the month before the start of our planning period—had been 225. We can find the amount of the change in the manpower level for April from the relationship

April Usage	–	March Usage	=	Change
$(.05x_{11} + .07x_{21})$	–	225	=	Change

Note that the change can be positive or negative. A positive change reflects an increase in the man-hour level, and a negative change reflects a decrease. Using this relationship, we can now specify the following constraint for the change in the number of man-hours used in April.

$$(.05x_{11} + .07x_{21}) - 225 = I_1 - D_1$$

Of course we cannot have an increase and decrease in the same period so either I_1 or D_1 will be zero. If April requires 245 man-hours, $I_1 = 20$ and $D_1 = 0$. If April requires only 175 man-hours, $I_1 = 0$ and $-D_1 = -50$; therefore $D_1 = 50$. This technique of denoting the change in man-hour requirements as the difference of two variables (I_m and D_m) means that, even though these variables will both be forced to assume nonnegative values by our linear programming model, we can represent both positive and negative fluctuations. Using the same approach in the following months (always subtracting the previous month's manpower levels from the current month's), we have the following constraints for the second and third months:

$$(.05x_{12} + .07x_{22}) - (.05x_{11} + .07x_{21}) = I_2 - D_2$$
$$(.05x_{13} + .07x_{23}) - (.05x_{12} + .07x_{22}) = I_3 - D_3$$

Placing the variables on the left-hand side and the constants on the right-hand side, the complete set of manpower smoothing constraints can be written as

$$
\begin{aligned}
.05x_{11} + .07x_{21} &&&& - I_1 + D_1 &= 225 \\
-.05x_{11} - .07x_{21} + .05x_{12} + .07x_{22} &&&& - I_2 + D_2 &= 0 \\
-.05x_{12} - .07x_{22} + .05x_{13} + .07x_{23} - I_3 + D_3 &= 0.
\end{aligned}
$$

Our initially rather small 2-product, 3-month scheduling problem has now developed into an 18-variable, 20-constraint linear programming problem. Notice that in our problem we were only concerned with one type of machine process, one type of manpower, and one type of storage area. In actual production scheduling problems you may encounter several machine types, several labor grades, and/or several storage areas. Thus, you are probably beginning to realize how large-scale linear programs of production systems come about. The large-scale model discussed in Chapter 6 has 500 products, 12 months, and 50 machine centers. Thousands of variables and thousands of constraints are necessary for the complete formulation.

When we encounter a large-scale system, we look for ways to reduce the number of variables and number of constraints so that our linear programming problem will be at least a little easier to solve. Some tips on doing this are provided in Chapter 6. For the Bollinger Electronics Company production scheduling problem presented in this section, we can reduce the number of variables by noting the following:

$$\begin{pmatrix} \text{Ending} \\ \text{Inventory} \\ \text{for a Given} \\ \text{Month} \end{pmatrix} = \begin{pmatrix} \text{Inventory} \\ \text{at Begin-} \\ \text{ning of the} \\ \text{Planning} \\ \text{Period} \end{pmatrix} + \begin{pmatrix} \text{All Produc-} \\ \text{tion up to} \\ \text{\& Including} \\ \text{the Current} \\ \text{Month} \end{pmatrix} - \begin{pmatrix} \text{All Ship-} \\ \text{ments or} \\ \text{Demands Up} \\ \text{to \& In-} \\ \text{cluding the} \\ \text{Current} \\ \text{Month} \end{pmatrix}.$$

Thus for component 322A we have:

$$s_{11} = 500 + x_{11} \qquad\qquad -1{,}000 \qquad\qquad = x_{11} \qquad\qquad -\ 500$$
$$s_{12} = 500 + x_{11} + x_{12} \qquad -1{,}000 - 3{,}000 \qquad = x_{11} + x_{12} \qquad -3{,}500$$
$$s_{13} = 500 + x_{11} + x_{12} + x_{13} - 1{,}000 - 3{,}000 - 5{,}000 = x_{11} + x_{12} + x_{13} - 8{,}500$$

Similarly, for component 802B we have:

$$s_{21} = x_{21} \qquad\qquad -\ 800$$
$$s_{22} = x_{21} + x_{22} \qquad -1{,}300$$
$$s_{23} = x_{21} + x_{22} + x_{23} - 4{,}300$$

We can now return to the objective function and constraints of our original linear programming formulation, and for every inventory variable (s_{im}) we can substitute the appropriate expression from above. What have we accomplished? We can now write the objective function and constraints with only the x_{im}, I_m, and D_m variables, and our problem is reduced to a 12-variable, 20-constraint problem. We have reduced the number of variables by 33%. In large-scale systems a reduction in variables of this magnitude can be quite significant.

The complete solution to the Bollinger Electronics Company production scheduling problem is shown in Table 5–L.

Production Schedule	April	May	June
Component 322A	500	3,050	5,350
Component 802B	2,858	1,642	–
Man-Hours	225	267.5	267.5
Inventory Schedule			
Component 322A	–	50	400
Component 802B	2,058	3,200	200

Total Cost = $224,378

(Includes Production, Inventory, and Manpower Smoothing Costs)

Table 5–L. Optimal Production, Manpower and Inventory Policy for Bollinger Electronics Company

At first glance the variation in the production schedule may look rather strange. But let us examine the logic of the recommended solution. Recall that the inventory cost for component 802B is one-half the inventory cost for component 322A. Therefore, as might be expected, component 802B tends to be carried in inventory, while the more expensive component 322A tends to be produced when demanded.

Why do we recommend producing over 2800 units of 802B in April when at least some of the units cannot be shipped until June? The answer to this is also logical. Recall that a manpower level of 225 man-hours was used in March. The low demand in April tends to dictate a manpower cutback; however, as you can see from the May and June demands, the firm would then have to rehire or add manpower in later months. The model is in effect smoothing the manpower requirements. Rather than recommending expensive manpower fluctuations, the linear programming model indicates it is cheaper to maintain a relatively high April production even though it means a higher inventory cost for component 802B. Keeping the April labor force at 225 man-hours means the only labor force change will be a 42.5 man-hour increase in May. This level will be maintained during the month of June.

We have seen in this illustration that a linear programming application of a relatively small two-product system (12 variables and 20 constraints) has provided some valuable information in terms of identifying a minimum-cost production schedule. In larger systems where the number of variables and constraints are too numerous to humanly track, linear programming models often provide significant cost savings for the firm.

Manpower Planning

Manpower planning or scheduling problems frequently occur when managers must make decisions involving departmental staffing requirements for a given period of time. This is particularly true when manpower assignments have some flexibility and at least some manpower can be assigned to more than one department or work center; this is often the case when employees have been cross-trained on two or more jobs. In the following example, we present a product mix problem similar to the Par, Inc. problem and show how linear programming can be used to determine not only an optimal product mix but also an optimal manpower allocation for the various departments.

McCarthy's Everyday Glass Company is planning to produce two styles of drinking glasses during the next month. The glasses are processed in four separate departments. Excess equipment capacity is available and will not be a constraining factor; however, the company's manpower resources are limited and will probably constrain the production volume for the two products. The man-hour requirements per case produced (one dozen glasses) are shown in Table 5–M.

Department	Product 1	Product 2
1	.070	.100
2	.050	.084
3	.100	.067
4	.010	.025

Table 5–M. Man-hours of Labor per Case of Product

The company makes a profit of $1.00 per case of product 1 and $.90 per case of product 2. If the number of man-hours available in each department is fixed, we can formulate Mc-

Carthy's problem as a standard product mix linear program. We use the usual notation:

x_1 = cases of product 1 manufactured

x_2 = cases of product 2 manufactured

b_i = man-hours available in department i, i= 1, 2, 3, 4.

The linear program can be written as

maximize $1.00x_1 + .90x_2$

s.t.

$$.070x_1 + .100x_2 \leq b_1$$
$$.050x_1 + .084x_2 \leq b_2$$
$$.100x_1 + .067x_2 \leq b_3$$
$$.010x_1 + .025x_2 \leq b_4$$
$$x_1, x_2 \geq 0.$$

To solve the normal product mix problem we would ask the production manager to specify the man-hours available in each department (b_1, b_2, b_3, and b_4); then we could solve for the profit maximizing product mix. However, in this case we assume that the manager has some flexibility in allocating manpower resources, and we would like to make a recommendation for this allocation as well as determining the optimal product mix.

Suppose that after consideration of the training and experience qualifications of the workers, we find this additional information:

Possible Labor Assignments	Man-hours Available
Dept. 1 only	430
Dept. 2 only	400
Dept. 3 only	500
Dept. 4 only	135
Depts. 1 or 2	570
Depts. 3 or 4	300
Total	2335

Of the 2335 man-hours available for the month's production, we see that 870 man-hours can be allocated with some management discretion. The constraints for the man-hours available per department are as follows:

$$b_1 \leq 430 + 570 = 1000$$
$$b_2 \leq 400 + 570 = 970$$
$$b_3 \leq 500 + 300 = 800$$
$$b_4 \leq 135 + 300 = 435$$

Since the 570 man-hours that have a flexible assignment between departments 1 and 2 cannot be assigned to both departments, we need the following additional constraint:

$$b_1 + b_2 \leq 430 + 400 + 570 = 1400.$$

Similarly, for the 300 man-hours that can be allocated between departments 3 and 4, we need the constraint:

$$b_3 + b_4 \leq 500 + 135 + 300 = 935.$$

In this formulation we are now treating the manpower assignments to departments as variables. The objective function coefficients for these variables will be zero since the b_i variables do not directly affect profit. Thus, placing all variables on the left-hand side of the constraints, we have the following complete formulation:

$$\text{maximize } 1.00x_1 + .90x_2 + 0b_1 + 0b_2 + 0b_3 + 0b_4$$

s.t.

$$
\begin{aligned}
.070x_1 + .100x_2 - b_1 & & & & \leq & 0 \\
.050x_1 + .084x_2 \phantom{{}- b_1} - b_2 & & & & \leq & 0 \\
.100x_1 + .067x_2 \phantom{{}- b_1 - b_2} - b_3 & & & \leq & 0 \\
.010x_1 + .025x_2 \phantom{{}- b_1 - b_2 - b_3} - b_4 & & \leq & 0 \\
b_1 & & \leq & 1000 \\
b_2 & & \leq & 970 \\
b_3 & & \leq & 800 \\
b_4 & \leq & 435 \\
b_1 + b_2 & \leq & 1400 \\
b_3 + b_4 & \leq & 935
\end{aligned}
$$

$$x_1, x_2, b_1, b_2, b_3, b_4 \geq 0.$$

This linear programming model will actually solve two problems: (1) it will find the optimal product mix for the planning period, and (2) it will allocate manpower to the departments in such a fashion that profits will be maximized. The solution to this six-variable, ten-constraint model is shown in Table 5–N.

Production Plan

Product 1 (x_1) = 4700 cases
Product 2 (x_2) = 4543 cases

Manpower Plan

Department 1	783 hours
Department 2	617 hours
Department 3	774 hours
Department 4	161 hours
Total	2335 hours

Profit = \$8,789

Table 5–N.   Optimal Production and Manpower Plan for McCarthy's Everyday Glass Company

Note that the optimal manpower plan utilizes all 2335 man-hours of labor by making the most profitable manpower allocations. In this particular solution there is no idle time in any of the departments. While this will not always be the case in problems of this type, if the manager does have the freedom to assign certain employees to different departments, the effect will probably be a reduction in the overall idle time. The linear programming model automatically assigns such employees to the departments in the most profitable manner. If the manager had used his judgment to allocate the man-hours to the departments, and we had then solved the product mix problem with fixed b_i's, we would in all probability have found slack in some departments while other departments represented bottlenecks because of insufficient resources.

Variations in the basic formulation of this section might be used in situations such as allocating raw material resources to products, allocating machine time to products, and allocating salesforce time to product lines or sales territories.

5.4   The Accounting Point of View

Most linear programming applications in business and industry are probably more readily identified as finance, marketing, or management oriented problems.   However, if you consider the data requirements of these applications (costs, profit margins, etc.) you should begin to realize the importance of the accounting function in formulating linear programming models. Can you imagine trying to determine an optimal product mix without the accountant's analysis of material costs, labor costs, factory overhead, selling and administrative expenses and the resulting product profitability?

Even if the applications of linear programming are not solely accounting problems, the accountant must still be aware of the business applications of the technique.   Anytime any department, division, etc., of a firm undertakes a cost control or a profit maximization project, the firm's accountants will become involved in the problem.   The accountants must therefore be thoroughly familiar with the assumptions underlying linear programming approaches to these problems.   With this background the accountant can provide valuable assistance in the formulation and evaluation of linear programming models.

While we do not present a specific accounting application of linear programming in this section we feel that almost all linear programming problems are accounting related applications.   We mean this in the sense that much of the data necessary for linear programming applications are provided by the accountant.   Since the success of the application is critically dependent on the reliability of the data used, the accountant's role in linear programming applications can be quite significant.

5.5   Distribution and Scheduling Applications

The Transportation Problem

The transportation problem is probably the most widely known application of linear programming.   This classical problem is concerned with making decisions with respect to the selection of distribution routes between manufacturing plants and regional warehouses or between warehouses and other distribution outlets.   The problem, as we shall consider it, is a combination production management and marketing distribution problem. It requires the decision maker to determine how much each of several plants should produce and how much of this production each plant should ship to each distribution center.

We begin our analysis in this section by studying a specific transportation problem faced by Foster Generators, Inc. After the Foster Generator problem has been formulated and solved, a general transportation model will be developed. Some of the more common extensions of the transportation problem will then be discussed in this general context.

Foster Generators, Inc. is a firm that has production operations in the relatively small towns of Wellsboro, Pennsylvania, Bedford, Indiana, and Jackson, Tennessee. Suppose that for the next three-month planning period, the plant production capacities for one particular type of generator are as follows:

Plant	Three-Month Capacity (units)
Wellsboro	5000
Bedford	7000
Jackson	3000

Suppose further that the firm distributes the product through four regional distribution centers located in New York, Chicago, St. Louis, and Atlanta, and that the three-month forecasted demands are as follows:

Distribution Center	Forecasted Three-Month Demand (units)
New York	6,000
Chicago	4,000
St. Louis	2,000
Atlanta	1,500

Foster would like to determine how much it should produce at each of its plants and how much of this production should be shipped from each plant to each distribution center.

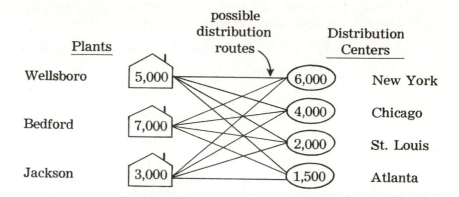

Question: Which routes to use and how much should be shipped on each?

Figure 5–B.   Possible Distribution Routes for the Foster Generators Transportation Problem

If we can assume the production costs are identical at the three plants, the only variable costs involved are transportation costs, and the problem becomes one of determining the distribution routes to be used and the quantity to be shipped via each route so that all distribution center demands can be met with a minimum total transportation cost. The cost for each unit shipped on each route is given in Table 5–O.

		To			
		New York	Chicago	St. Louis	Atlanta
	Wellsboro	2.80	6.00	7.25	5.75
From	Bedford	5.50	3.25	2.00	2.50
	Jackson	6.75	4.50	2.75	2.75

Table 5–O.   Transportation Cost/Unit for the Foster Generators Transportation Problem

We should now be able to formulate the linear programming model for this transportation problem. Letting

x_{ij} = amount shipped from plant i to distribution center j

the objective function becomes

$$\text{Minimize} \quad 2.80x_{11} + 6.00x_{12} + 7.25x_{13} + 5.75x_{14}$$
$$+ 5.50x_{21} + 3.25x_{22} + 2.00x_{23} + 2.50x_{24}$$
$$+ 6.75x_{31} + 4.50x_{32} + 2.75x_{33} + 2.75x_{34}.$$

The constraints affecting the problem are that all the demands must be met and that plant production capacities cannot be exceeded. We guarantee that demand will be satisfied by specifying that

$$x_{11} + x_{21} + x_{31} = 6000 \quad \text{New York}$$
$$x_{12} + x_{22} + x_{32} = 4000 \quad \text{Chicago}$$
$$x_{13} + x_{23} + x_{33} = 2000 \quad \text{St. Louis}$$
$$x_{14} + x_{24} + x_{34} = 1500 \quad \text{Atlanta}.$$

The satisfaction of the production capacity restrictions is guaranteed by the following constraints:

$$x_{11} + x_{12} + x_{13} + x_{14} \leq 5000 \quad \text{Wellsboro}$$
$$x_{21} + x_{22} + x_{23} + x_{24} \leq 7000 \quad \text{Bedford}$$
$$x_{31} + x_{32} + x_{33} + x_{34} \leq 3000 \quad \text{Jackson}.$$

In addition, we include the usual nonnegativity restrictions on each of the x_{ij}.

The solution to this twelve-variable, seven-constraint linear programming model provides a production and distribution plan for Foster Generators which minimizes total transportation cost. This solution is presented in Table 5–P and Figure 5–C.

		New York	Chicago	St. Louis	Atlanta
	Wellsboro	5,000			
From	Bedford	1,000	4,000	2,000	
	Jackson				1,500

Total Transportation Costs = $40,625

Table 5–P.   Minimum Cost Transportation Plan for Foster Generators, Inc.

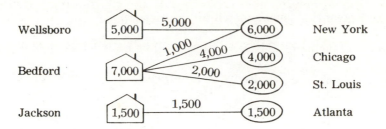

Figure 5–C. Optimal Transportation Schedule for Foster Generators, Inc.

We now turn to a development of the general transportation model. The general linear programming formulation of the m–plant, n–distribution center transportation model is

$$\text{minimize} \sum_{i=1}^{m} \sum_{j=1}^{n} c_{ij} x_{ij}$$

s.t.

$$\sum_{j=1}^{n} x_{ij} \leq S_i \quad i = 1, 2, \ldots, m \quad \text{(Supply)}$$

$$\sum_{i=1}^{m} x_{ij} = D_j \quad j = 1, 2, \ldots, n \quad \text{(Demand)}$$

$$x_{ij} \geq 0 \quad \text{for all i and j}$$

where c_{ij} = transportation cost/unit for items shipped from plant i to distribution center j, $i = 1, 2, \ldots, m$
$j = 1, 2, \ldots, n$

S_i = supply or capacity in units at plant i, $i = 1, 2, \ldots, m$

D_j = demand in units at distribution center j, j
$= 1, 2, \ldots, n.$

You might have noted in the Foster Generators problem that the total production capacity for the three plants (15,000 units) exceeded the total demand at the four distribution centers (13,-500 units). As long as the total supply in a transportation problem is at least equal to the total demand, a feasible solution will result and hence all demands will be satisfied. If we encounter a transportation problem where total supply ($\sum_{i=1}^{m} S_i$) is less than total demand ($\sum_{j=1}^{n} D_i$) there will be no feasible solution to the problem; i. e., we know in advance that demands cannot be satisfied unless $\sum_{i=1}^{m} S_i \geq \sum_{j=1}^{n} D_j.$ (5.1)

We can still generate a minimum transportation cost solution for the available supply and at the same time identify the destinations that will not receive the requested demand by creating a fictitious plant or source with a supply exactly equal to the difference between the total demand and total supply. If we let S_{m+1} indicate the added fictitious supply, then

$$S_{m+1} = \sum_{j=1}^{n} D_j - \sum_{i=1}^{m} S_i.$$

Essentially, we create a fictitious plant with a capacity of S_{m+1}.

The result of introducing this fictitious plant is to make total supply equal to total demand. Thus condition (5.1) holds, and the modified linear programming model will have a feasible solution. Since no shipments will actually be made from the fictitious plant, all objective function cost coefficients for this source can be set equal to zero. Thus,

$$c_{m+1, 1} = c_{m+1, 2} = \ . \ . \ . \ = c_{m+1, n} = 0.$$

The interpretation of fictitious plant variables occurring in the optimal solution is that the destinations receiving these shipments will have a shortage equal to the value of the fictitious variable. Thus in transportation problems where the supply is less than demand a model incorporating a fictitious plant will identify the minimum cost shipments for the units actually produced. It will also indicate the demand centers where the shortages will actually occur.

As another extension of the basic transportation model, we can consider cases in which some routes are physically unacceptable or have a limited shipping capacity. In the first case, suppose that Foster Generators, Inc. decided that under no circumstances would Foster ship from the Jackson plant to the New York distribution center. We can guarantee this result in our linear programming solution by removing variable x_{31} from the model or by assigning an arbitrarily large cost per unit for that route. For example, using this large cost per unit approach we might make $c_{31} = \$1,000$ per unit. Except in the case where the Jackson to New York route must be used to reach a feasible solution, the arbitrarily large cost will prevent shipment over this route. Other routes that are judged unacceptable, or highly undesirable, in advance can be handled in a similar manner.

In the case where routes are acceptable up to a limited shipping capacity, additional constraints will be necessary. For example, suppose that the Bedford to Chicago route had a maximum capacity of 3,000 units over the three-month period. The following constraint guarantees that the route capacity will be met:

$$x_{22} \leq 3000.$$

In general, if L_{ij} represents the route capacity from plant i to destination j, additional constraints of the form

$$x_{ij} \leq L_{ij}$$

will be needed.

For another variation of the basic transportation model, consider the case where production costs vary at the different plants. If this condition exists, we can simply redefine c_{ij} as the production cost per unit at plant i plus the unit transportation cost for items shipped from plant i to destination j. The optimal solution to this transportation model provides both a minimum cost production plan and a minimum cost distribution plan.

For reasonably sized problems, the solution to the general transportation problem is quite easily handled by solving it as a linear programming problem. However, several specialized computational algorithms have been devised specifically for this problem, (e. g., the Stepping Stone Method, MODI, etc.). Discussions of these algorithms can be found in many of the texts listed in the Bibliography. However, since the Simplex method solves most linear programming transportation models quite satisfactorily, we will not go into the details of these specialized transportation algorithms in this text.

The Assignment Problem

The assignment model is another general linear programming model that has been applied to a variety of administrative problems. Typical assignment problems involve assigning jobs to machine centers or plants, assigning workers to tasks or projects, assigning salesmen to sales territories, assigning contracts to bidders, etc. A distinguishing feature of this problem is that one job, worker, etc., is assigned to one and only one plant, project, etc. Typically, we are looking for the assignments that will optimize a stated objective such as cost, profit, time, etc.

As an illustration we consider the case of Fowle Marketing Research, Inc. which has just received requests for market research studies from three different clients. The company is faced with the task of assigning project leaders to each of these three research studies. Currently, four individuals are relatively free from other major commitments and are available for the project leader assignments. Fowle's management realizes that the time required to complete the studies will depend upon the experience and ability of the project leaders.

After considering all possible assignments, the responsible manager has made an estimate of the project completion times for each leader-client combination. This data is summarized in Table 5–Q. It is estimated, for example, that McClymonds would require six days to complete client A's project while Carle would require approximately nine days for the same project.

			Client		
			1.	2.	3.
			A	B	C
	1.	Terry	10	15	9
Project	2.	Carle	9	18	5
Leader	3.	McClymonds	6	14	3
	4.	Higley	12	16	8

Table 5–Q. Estimated Project Completion Time (in days) for the Fowle, Inc. Assignment Problem

Since the three projects have been judged to have approximately the same priority, the company would like to assign project leaders such that the total man-days required to complete the three projects would be a minimum. If a project leader can be assigned to at most one client, what assignments should be made?

Let us attempt a linear programming formulation of this assignment problem. Once again we use the double subscript notation for the decision variables.

$$
x_{ij} = \begin{cases} 1 & \text{if project leader } i \text{ is assigned to client } j, \\ & i = 1, 2, 3, 4; \ j = 1, 2, 3 \\ 0 & \text{otherwise} \end{cases}
$$

Wait a minute, you say! How can these variables be appropriate linear programming decision variables when they can only take on the values zero or one? Technically you are correct; they cannot be. Actually the problem we are solving is a zero-one integer programming problem. However, this problem has a very nice feature that makes it possible for us to treat it as a linear programming problem. That is, the values of all the basic variables will be either zero or one at the extreme points of the feasible region (see exercise 13). Therefore, since the optimal solution to a linear program lies at an extreme point, the optimal solution to a linear programming formulation of the assignment problem will have the basic variables all equal to zero or one. Thus we see that solving the linear programming formulation provides the optimal solution to the assignment problem.

Using the above decision variables, the objective function calling for the minimization of total man-days can be written as

$$
\begin{aligned}
\min \ & 10x_{11} + 15x_{12} + 9x_{13} \\
& + 9x_{21} + 18x_{22} + 5x_{23} \\
& + 6x_{31} + 14x_{32} + 3x_{33} \\
& + 12x_{41} + 16x_{42} + 8x_{43}.
\end{aligned}
$$

The constraints affecting this problem are that all clients must receive exactly one project leader and that the project leaders cannot be assigned to more than one client. The first condition is satisfied by the following linear constraints:

$$x_{11} + x_{21} + x_{31} + x_{41} = 1 \quad \text{Client A}$$
$$x_{12} + x_{22} + x_{32} + x_{42} = 1 \quad \text{Client B}$$
$$x_{13} + x_{23} + x_{33} + x_{43} = 1 \quad \text{Client C}.$$

Note that although $x_{11} = \frac{1}{2}$, $x_{21} = \frac{1}{2}$ is a feasible solution, it could never be optimal because it does not correspond to an extreme point. Hence, at the optimal solution these constraints do indeed guarantee that each client will receive exactly one project leader.

The second condition above is reflected in the following constraints:

$$x_{11} + x_{12} + x_{13} \leq 1 \quad \text{Terry}$$
$$x_{21} + x_{22} + x_{23} \leq 1 \quad \text{Carle}$$
$$x_{31} + x_{32} + x_{33} \leq 1 \quad \text{McClymonds}$$
$$x_{41} + x_{42} + x_{43} \leq 1 \quad \text{Higley}$$

In addition, the nonnegativity constraints for the variables are included as usual.

The solution to this twelve-variable, seven-constraint linear programming model provides a project leader assignment plan that will allow Fowle Company to complete the three projects with a minimum total number of man-days expended. This solution is shown in Table 5–R.

Client	Assigned Project Leader	Estimated Man-Days
A	McClymonds	6
B	Terry	15
C	Carle	5
	Total	26

Unassigned: Higley

Table 5–R.   Minimum Man-Day Assignment for Fowle Marketing Research, Inc.

The general assignment problem is one that involves m–persons and n–tasks. Introducing the notation

$$c_{ij} = \text{cost for assigning person i to task j,}$$
$$i = 1, 2, \ldots, m$$
$$j = 1, 2, \ldots, n$$

we can write the general assignment model as follows.

$$\text{minimize} \quad \sum_{i=1}^{m} \sum_{j=1}^{n} c_{ij} x_{ij}$$

s.t.

$$\sum_{j=1}^{n} x_{ij} \leq 1 \quad i = 1, 2, \ldots, m \quad \text{Persons Available}$$

$$\sum_{i=1}^{m} x_{ij} = 1 \quad j = 1, 2, \ldots, n \quad \text{Task Demands}$$

$$x_{ij} \geq 0 \quad \text{for all i and j}$$

By looking back to the general transportation model in the previous section you can see that the assignment problem is a special case of the transportation problem. Specifically, it is the case where the supplies at all origins are one ($S_i = 1$ for all i) and the demands at all destinations are also one ($D_j = 1$ for all j).

Like the transportation model, the above assignment model will only have a feasible solution if the total supply (m) is at least as great as the total demand for assignments (n). If the demand for assignments exceeds the supply, we can add fictitious origins until m = n and a feasible solution exists. For example, if Fowle, Inc. had had five clients and only three project leaders available, we would have had to add two fictitious project leaders to obtain a feasible linear programming solution. In the problem formulation, the cost for assigning the fictitious leaders would be zero (i. e., $c_{ij} = 0$ for all fictitious project leaders). If this approach was employed, projects receiving a fictitious project leader could not be started until a project leader became available sometime in the future.

If some assignments are unacceptable (for example, one person is not qualified to handle a given client), we can omit this

x_{ij} variable from the model or assign an arbitrarily large cost to the corresponding cost coefficient c_{ij} in the objective function. Recall that this is the identical procedure used to handle unacceptable routes in a transportation problem.

As a final note, assignment problems like transportation problems have special purpose algorithms that can be used to find optimal solutions. Such algorithms tend to be somewhat more computationally efficient, especially for large assignment problems. However, since the solution by the Simplex method is quite satisfactory, we will not attempt to show the details of these special purpose algorithms in this text.

5.6 Ingredient Mix Applications

The Blending Problem

Blending problems arise whenever a manager must decide how to blend two or more resources in order to produce one or more products. In these situations, the resources contain one or more essential ingredients that must be blended in such a manner that the final products will contain specific percentages of the essential ingredients. In most of these applications, then, management must decide how much of each resource to purchase in order to satisfy product specifications and product demands at minimum cost.

These types of problems occur frequently in the petroleum industry (e. g., blending crude oil to produce different octane gasolines), chemical industry (e. g., blending chemicals to produce fertilizers, weed killers, etc.), and food industry (e. g., blending input ingredients to produce soft drinks, soups, etc.). Because of their widespread application, our objective in this section is to illustrate how linear programming can be applied to solve these types of problems.

Consider the case of Beauty Suds, Inc., manufacturers of Wonderful Hair Shampoo. Beauty Suds is considering the production of a new product: Wonderful Plus Hair Shampoo. The new product is a blend of the company's standard shampoo base product, a new dandruff preventive, a perfume, and deionized water. The company has specified the following final product characteristics per gallon manufactured:

	Minimum	Maximum
Suds forming ingredient	100 grams	150 grams
Dandruff ingredient	50 grams	50 grams
Perfume ingredient	20 grams	30 grams
Shampoo viscosity in centipose	400 c.p.	600 c.p.

The cost and general characteristics of the four raw materials are as follows:

	Shampoo Base	Dandruff Preventive	Perfume	Deionized Water
Suds ingredient (gr/gal)	150	0	0	0
Dandruff ingredient (gr/gal)	10	500	0	0
Perfume ingredient (gr/gal)	15	0	200	0
Viscosity (c.p.)	700	600	400	5
Cost/gal.	$3.00	$15.00	$60.00	$.25

Assuming that all quantities of ingredients blend linearly by volume, the management of Beauty Suds would like to know how much of each raw material should be in each gallon of the new shampoo product in order to meet product requirements at minimum cost.

In order to formulate a linear programming model for the Beauty Suds blending problem, we begin by defining appropriate decision variables. Let

x_1 = gallons of standard shampoo base per gallon of shampoo,

x_2 = gallons of dandruff preventive per gallon of shampoo,

x_3 = gallons of perfume per gallon of shampoo, and

x_4 = gallons of deionized water per gallon of shampoo.

The objective function for our problem can then be written as follows:

$$\min \ 3.00x_1 + 15.00x_2 + 60.00x_3 + 0.25x_4.$$

To meet the requirements for the minimum and maximum amounts of suds forming ingredient, dandruff ingredient, perfume ingredient, and shampoo viscosity, we formulate the following set of constraints:

$$
\begin{array}{lll}
150x_1 & \leq 150 \left.\vphantom{\begin{matrix}a\\b\end{matrix}}\right\} & \text{Suds} \\
150x_1 & \geq 100 \\
10x_1 + 500x_2 & = 50 & \text{Dandruff} \\
15x_1 \qquad + 200x_3 & \leq 30 \left.\vphantom{\begin{matrix}a\\b\end{matrix}}\right\} & \text{Perfume} \\
15x_1 \qquad + 200x_3 & \geq 20 \\
700x_1 + 600x_2 + 400x_3 + 5x_4 & \geq 400 \left.\vphantom{\begin{matrix}a\\b\end{matrix}}\right\} & \text{Viscosity} \\
700x_1 + 600x_2 + 400x_3 + 5x_4 & \leq 600
\end{array}
$$

In addition, to guarantee that the amount of raw material blended will produce exactly one gallon of Wonderful Plus Hair Shampoo, we require that

$$x_1 + x_2 + x_3 + x_4 = 1.$$

At the optimal solution then, if management requires 1000 gallons of the new product they need only multiply the values of the decision variables by 1000. (Equivalently, we could have replaced the right-hand side of this constraint by 1000 and realized exactly the same solution.)

After adding the usual nonnegativity requirements, the complete linear programming model was solved using the Simplex method. The optimal solution is shown in Table 5–S.

Material	Quantity (gallons)
Shampoo Base	.759
Dandruff Preventive	.085
Perfume	.043
Water	.113

Cost per Gallon = $6.16

Table 5–S. Optimal Solution for the Beauty Suds, Inc. Blending Problem

We see that the optimal solution to the problem tells management not only how to blend the available resources to meet product specifications, but also provides management with the cost of doing so. This additional information is often the most critical piece of information from management's point of view. For example, suppose a priori that management felt that it could not market the product unless the cost of raw materials for the new product was less than $7.00 per gallon. Given this a priori assessment should management produce the new product? Clearly the answer is yes, since the linear programming solution to the problem shows management that the raw materials cost to produce the new product is $6.16 per gallon.

In the Beauty Suds blending problem we formulated as a linear programming model, we saw a problem situation wherein four different resources were blended to produce one product. In many blending problems, however, a number of different products must be produced. A descriptive approach to defining appropriate decision variables for these more general blending problems is to use double subscripted decision variables; the first subscript can be used to denote the resource and the second subscript used to denote the product. Thus, we let

$$x_{ij} = \text{amount (e. g., gallons) of resource i used to produce product j.}$$

Problem 4 at the end of the Chapter describes a gasoline blending situation where decision variables of the above type can be applied.

The Diet Problem

The diet problem is presented in this chapter in order to introduce the reader to this well known application of linear programming. Typically the diet problem, or in agricultural applications the feed-mix problem, involves specifying a food or feed ingredient combination that will satisfy some minimal nutritional requirements at a minimum total cost. As a result, some authors view the diet problem as a special case of the general blending problem.

Let us consider the feed-mix form of the diet problem encountered by Bluegrass Farms, Inc. in Lexington, Kentucky. This company is experimenting with a special diet for its race

horses.   The feed components available for the diet are a standard horse feed product, a vitamin enriched oat product, and a
new vitamin and mineral feed additive.   Table 5–T shows the
nutritional values and costs for the three components.

| | | Feed Component | | |
		Standard	Enriched Oats	Additive
Diet	Ingredient A	.8	.2	0
Requirement	Ingredient B	1.0	1.5	3.0
	Ingredient C	.1	.6	2.0
	Cost/lb.	$ .25	$ .50	$3.00

Table 5–T.   Units of Feed Ingredient per Pound of Feed Component

Suppose that the horse trainer sets the minimum daily diet
requirement at 3 units of ingredient A, 6 units of ingredient B,
and 4 units of ingredient C.   Also suppose that for weight control, the trainer does not want the total daily feed to exceed
six pounds.   What is the optimal daily mix of the three components?

A linear programming model of this diet problem can be formulated as follows.   Let

$x_1 =$ pounds of the standard feed,

$x_2 =$ pounds of the enriched oats, and

$x_3 =$ pounds of the additive.

With the overall goal of minimizing cost, the objective function
becomes

$$\min .25x_1 + .50x_2 + 3.00x_3$$

Using the information provided, the constraints for our problem
are easily formulated as follows:

$$.8x_1 + .2x_2 \qquad\qquad \geq 3 \quad \text{Ingredient A}$$
$$1.0x_1 + 1.5x_2 + 3.0x_3 \geq 6 \quad \text{Ingredient B}$$
$$.1x_1 + .6x_2 + 2.0x_3 \geq 4 \quad \text{Ingredient C}$$
$$x_1 + x_2 + x_3 \leq 6 \quad \text{(Total weight)}$$
$$x_1, x_2, x_3 \geq 0 .$$

The minimum cost feed-mix solution for the above linear programming model is given in Table 5–U.

Standard Feed	3.51 pounds
Enriched Oats	.95
Vitamin Additive	1.54
Total	6.00 pounds

Daily Cost = $5.97

Table 5–U. Minimum Cost Diet for Bluegrass Farms, Inc. Horses

As we saw in the blending problem, the optimal solution for the Bluegrass Farms linear programming model not only tells management how to mix the three components to produce the desired product, but also the cost of doing so. This latter piece of information is often the first thing management wants to know. For example, if Bluegrass Farms is presently purchasing a feed mix with similar characteristics for a daily cost of less than $5.97, it is doubtful whether they would consider producing this new special diet except on an experimental basis.

5.7 Environmental Protection Application

While linear programming has been applied primarily in business and industrial settings, the technique is of course not limited to these fields. Applications of linear programming to health care, environmental protection, and a variety of other problems society is currently faced with have been made. In this section we describe a problem that, although similar to some of the industrial problems we have studied in earlier sections, incorporates environmental considerations. Specifically, the linear programming model we present will assist a firm in making policy decisions of an anti-pollution nature.

We consider the problem faced by Skillings Industrial Chemicals, Inc., a refinery located in southwestern Ohio near the Ohio River. The company's major product is manufactured from a chemical process that requires two raw materials, A and B. The production of one pound of product requires one pound of material A and two pounds of material B. The output of this process also yields one pound of liquid waste material and a solid waste byproduct. The solid waste byproduct is given to a local fertilizer plant as payment for picking up the byproduct. Since the liquid waste material has virtually no market value, and since it is in liquid form, the refinery has been dumping it directly into the Ohio River. Skillings' manufacturing process is shown schematically in Figure 5–D.

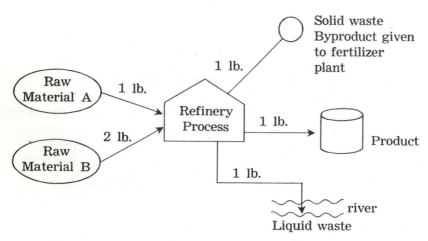

Figure 5–D.   Manufacturing Process at Skillings Industrial Chemicals, Inc.

Recent governmental pollution guidelines established by the Environmental Protection Agency will not permit this liquid waste disposal process to continue. Hence, the refinery's research group has developed the following set of alternative uses for the waste material:

1. produce a secondary product (K) by adding one pound of raw material A to every pound of liquid waste;

2. produce another secondary product (M) by adding one pound of raw material B to every pound of liquid waste;

3. specially treat the liquid waste so that it meets pollution standards before dumping directly into the river.

These three alternatives are depicted in Figure 5–E.

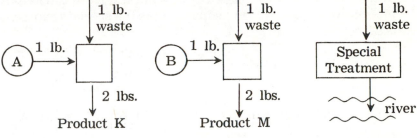

Figure 5–E.   Alternatives for Handling the Refinery Liquid
Waste Material

The company's management knows the secondary products
will be low in quality and will probably not be very profitable.
Management is also aware of the fact that the special treatment
alternative will be a relatively expensive operation.   The company's problem is to determine how to satisfy the pollution regu-
lations and still maintain the highest possible profit.   How
should the waste material be handled?   Should Skillings produce
product K, product M, use the special treatment, or employ some
combination of the three alternatives?

Since the waste disposal process will affect the production of
the firm's primary product, the complete system—manufacturing
process and waste disposal process—will have to be considered
together in the analysis.   Hopefully, Skillings will be able to
satisfy the anti-pollution requirements and still make a satisfactory profit.

Considering the selling price, material costs, and manpower
costs the accounting department has prepared the following information with respect to product profit contribution.

Product	Profit Contribution
Primary	$2.10/pound
Product K	−.10/pound
Product M	.15/pound

As you can see, the primary product is very profitable, while the secondary products are marginal. In fact, product K can only be produced at a loss. However, since product K provides a means for disposing of the waste material, it must still be considered as an alternative. Furthermore, suppose that the special treatment disposal cost is $.25 per pound.

The ingredients required to make one pound of each product are summarized in Table 5–V.

Ingredients	Primary Product	Product K	Product M
Raw Material A	1	0.5	0.0
Raw Material B	2	0.0	0.5
Waste	0	0.5	0.5

Table 5–V. Pounds of Ingredient Required
per Pound of Product

Additional restrictions on our problem result from the fact that during any planning period the company will have limited amounts of raw materials available. During the production period of interest in our current problem, these maximums are 5,000 pounds of material A and 7,000 pounds of material B.

Let us see how we can formulate a linear programming model that will help management solve this problem. We let

$x_1 =$ pounds of primary product,

$x_2 =$ pounds of secondary product K,

$x_3 =$ pounds of secondary product M, and

$x_4 =$ pounds of liquid waste material processed by the special treatment.

Assuming the liquid waste material is a zero-cost, zero-profit by-product of the primary process, it will only incur additional cost if it has to be specially treated. Thus the objective function can be written as

$$\max 2.10x_1 - .10x_2 + .15x_3 - .25x_4.$$

The raw material constraints are

$$x_1 + .5x_2 \qquad \leq 5000$$
$$2x_1 \qquad + .5x_3 \leq 7000$$

We note that the production of both product K (x_2) and product M (x_3) depends upon the amount of liquid waste material available. Hence, we must include a constraint on the amount of products x_2, x_3 and x_4 that can be produced. Since all the liquid waste material must be disposed of, we must require that

$$.5x_2 + .5x_3 + x_4 = \text{total liquid waste material available.}$$

Since the amount of total liquid waste material generated is equal to the amount of primary product produced (see Figure 5–D), we can write the above requirement as

$$.5x_2 + .5x_3 + x_4 = x_1,$$

or

$$-x_1 + .5x_2 + .5x_3 + x_4 = 0.$$

The solution to this four-variable, three-constraint linear programming model provides the most profitable production and pollution control plan. The complete solution is shown in Table 5–W.

	Production	Profit
Primary Product	3,500 pounds	$ 7350
Waste Disposal		
Product K	3,000 pounds	−$300
Specially Treated Waste	2,000 pounds	−$500
	Total Profit	$ 6550

Table 5–W.   Optimal Production and Waste Management Plan for Skillings, Inc.

We see that the optimal solution to our linear programming model involves the production of product K and specially treated waste, both of which result in losses for the company. Does this seem reasonable in view of the fact that product M will enable Skillings to dispose of waste and still realize some contribution to profit? Let's see if we can answer this question by analyzing the optimal solution a bit more carefully.

In our model, the primary product was so profitable when compared to the alternatives, we produced as much primary product as possible. Since B was the limiting raw material resource, all of B was used up in the production of primary product. Thus, since product M required raw material B, it was impossible to produce any amount of product M. Hence, the waste material generated had to be disposed of using product K and the specially treated waste process.

5.8   Summary

In this chapter we have tried to give the reader an appreciation and understanding of the broad range of situations in which linear programming may be a useful decision-making aid. Through illustrative situations, we have formulated and solved problems from the functional areas of finance, marketing, and management. We pointed out how the general cost control and profit maximization interests of the accounting function also makes the accountant an important contributor to the formulation and solution of these problems. In addition, we showed how linear programming could be applied to the classical transportation, assignment, blending, and diet problems. An application involving environmental protection showed the flexibility of linear programming in solving social problems.

All the illustrations presented in this chapter have been simplified versions of actual situations in which linear programming has been applied. In real-world applications, the reader will find that the problem is not as concisely stated, that the data is not as readily available, and that the problem has a larger number of variables and constraints. However, a thorough study of the applications in this chapter is a good place for the reader who eventually hopes to apply linear programming to real-world problems to begin. In the following chapter you will be exposed to the implementation problems that are often encountered in an actual large-scale application of linear programming.

For the reader interested in learning about additional applications of linear programming, we suggest a text by Gass.[1] In this text the author presents a bibliography of over one-hundred business and industrial applications of the technique.

---

[1] Gass, Saul J., *An Illustrated Guide to Linear Programming*, New York McGraw-Hill Book Co., 1970.

5.9 Problems

1. *Product mix*—A small job shop has purchased a new drill press which can be operated 40 hours per week. Two products are to be manufactured with this equipment. Product 1, which has a profit of 30 cents per unit, can be produced at the rate of 50 units per hour, and product 2, which has a profit of 50 cents per unit, can be produced at the rate of 40 units per hour. Based on current orders, 1000 units of product 1 and 500 units of product 2 must be manufactured each week.

   a. If the firm can sell all the units it produces, how many units of each product should be manufactured each week?

   b. What is the profit contribution of an overtime hour for the drill press? Is there an upper limit on the amount of overtime you would want? Explain.

2. *Investment planning*—The management of the Bordon Investment Company has three investment opportunities to consider over the next eighteen months. The investments differ in terms of availability date, duration, rate of return and maximum dollar amount. The data as summarized by one of the firm's financial analysts is as follows:

Investment	Available	Duration	Projected Annual Rate of Return	Max. Amount
Mutual Fund A	now	no limit	.09	no limit
Bond B	now	12 months	.12	$50,000
Stock C	6 months	no limit	.14	$25,000.

   The firm can buy or sell the mutual fund or stock any time, but if the bond investment is made, the firm must keep it for the full year.

   a. If the firm has $60,000 to invest over the next eighteen months, develop the investment plan for each six-month period that will maximize the return. Note that since investment decisions are made each six months, the rate of return over each six-month period will be one-half of the annual rate. Assume that the total funds available to the firm at the start of period two are the original $60,000 plus any 6-month interest from the mutual fund and that the total funds available at the start of period three are the $60,000 plus interest from all previous investments.

b. If the company changed the upper limit on the stock to a maximum of $50,000, what would happen? Use the dual variable to help you answer this before you resolve the problem.

3. *Make-or-buy*—The Carson Stapler Manufacturing Company forecasts a 5000 unit demand for its Sure-Hold model during the next quarter. This stapler is assembled from three major components: base, staple cartridge and handle. Previously, Carson has manufactured all three components; however, the 5000 forecasted units is a new high in sales volume, and it is doubtful that the firm will have production capacity to make all the components. The company is considering contracting a local firm to produce at least some of the components.

The production time requirements per unit are as follows:

Departments	Base	Cartridge	Handle	Total Dept. Time Available
A	.03 hrs.	.02 hrs.	.05 hrs.	400
B	.04 hrs.	.02 hrs.	.04 hrs.	400
C	.02 hrs.	.03 hrs.	.01 hrs.	400

After considering the firm's overhead, material, and labor costs the accounting department has determined the unit manufacturing cost for each component. This data along with the purchase price quotations by the contracting firm are as follows:

Component	Manufacturing Cost	Purchase Cost
Base	$ .75	$ .95
Cartridge	$ .40	$ .55
Handle	$1.10	$1.40

a. Determine the make-or-buy decision for Carson that will meet the 5000 unit demand at a minimum total cost. How many units of each component should be made and how many purchased?

b. Which departments are limiting the manufacturing volume? If overtime could be considered at the additional cost of $3.00 per hour, which department(s) should be allocated the overtime? Explain.

c. Suppose up to 80 hours of overtime can be scheduled in department A. What do you recommend?

4. *Blending problem*—Seastrand Oil Company produces two grades of gasoline: regular and high octane. Both types of gasoline can be produced from two types of crude oil. Although both types of crude oil contain the two important compounds required to produce both gasolines, the percentage of important compounds in each type of crude oil differs, as well as the cost per gallon. The composition of each type of crude oil and the cost per gallon is shown below.

Type of Crude Oil	Cost	Compound c_1	Compound c_2
1	$0.10	20%	60%
2	$0.15	50%	30%

Daily demand for regular octane gasoline is 800,000 gallons, and daily demand for high octane is 500,000 gallons. Each gallon of regular must contain at least 40% of c_1, whereas each gallon of high octane can contain at most 50% of c_2. How many gallons of each type of crude oil should Seastrand Oil purchase in order to satisfy daily demand at minimum cost?

5. *Marketing strategy*—In the marketing strategy problem encountered by Electronic Communications, Inc. (see section 5.2) we found the advertising budget and salesforce availability were the limiting constraints. Assume an additional $2,000 is available for advertising and an additional 300 hours is available in salesforce effort, how does the optimal solution change? Formulate and solve the dual. What do the dual variables now tell us?

6. *Paper trim problem*—The Ferguson Paper Company pro-
    duces rolls of paper for use in adding machines, desk cal-
    culators, and cash registers. The rolls, which are 200 feet
    long, are produced in widths of 1½, 2½, and 3½ inches. The
    production process provides 200-feet rolls in 10-inch widths
    only. The firm must therefore cut the rolls to the desired
    final product sizes. The seven cutting alternatives and the
    amount of waste of each are as follows:

Cutting Alternative	Number of 1½	2½	3½	Waste Inches
1	6	0	0	1
2	0	4	0	0
3	2	0	2	0
4	0	1	2	½
5	1	3	0	1
6	1	2	1	0
7	4	0	1	½

The minimum production requirements for the three products are as follows:

Roll Width	Units
1½	1,000
2½	2,000
3½	4,000

a. If the company wants to minimize the number of units
    of the 10-inch rolls that must be manufactured, how
    many units will be processed on each cutting alterna-
    tive? How many rolls are required, and what is the
    total inches of waste?

b. If the company wants to minimize the waste generated,
    how many 10-inch units will be processed on each cut-
    ting alternative? How many rolls are required, and
    what is the total inches of waste?

c. What are the differences in approaches (a) and (b) to
    this trim problem? In this case, which objective do you
    prefer? Explain. What are the types of situations that
    would make the other objective the more desirable?

7. *Inspection*—The Get-Well Pill Company inspects capsule medicine products by passing the capsules over a special lighting table where inspectors visually check for cracked or partially filled capsules. Currently any of three inspectors can be assigned to the visual inspection task. The inspectors, however, differ in accuracy and speed abilities and are paid at slightly different wage rates. The differences are as follows:

Inspector	Speed	Accuracy	Wage Rate
Davis	300/hr.	98%	2.95/hr.
Wilson	200/hr.	99%	2.60/hr.
Lawson	350/hr.	96%	2.75/hr.

Operating on a full eight-hour shift, the company needs at least 2000 capsules inspected with no more than 2% of these capsules having inspection errors. In addition, because of the fatigue factor of this inspection process, no one inspector can be assigned this task for more than four hours per day. How many hours should each inspector be assigned to the capsule inspection process during an eight-hour day? What volume will be inspected per day and what is the daily capsule inspection cost?

8. *Manpower assignments*—Manager Sparky Gibson of the Hamilton White Sox baseball team is trying to establish his starting pitchers for the crucial three-game series with the Mt. Washington Tigers. Sparky has the following five pitchers available.

Minta—the ace of the staff who just pitched last night's extra inning game against the Northtown Giants;

O'Donnel—the aging veteran who has 3 wins and 6 losses this season;

Banks—the relief ace who has started only one game this season;

Hudlow—the rookie who just arrived from the Delphi farm-team;

Nash—a 10-win, 4-loss right-hander who has been having arm problems.

Sparky knows the Tigers are saving their ace pitcher for the third game of the series. After considering the Tigers' probable line-up and pitchers, Sparky has estimated the probability of winning each of the three games with each of the five starting pitcher alternatives.

The winning probabilities are as follows:

		Game 1	Game 2	Game 3
	Minta	.60	.75	.65
	O'Donnel	.40	.45	.45
Starting	Banks	.50	.45	.35
Pitcher	Hudlow	.30	.50	.20
	Nash	.40	.45	.30

a. Assuming each pitcher could only start one game in the series, what is the pitching rotation that will provide the highest winning probability for the Sox?

b. If Nash reports that his arm is fine prior to the start of the first game and his probabilities of winning are revised to .60, .70, and .50 for the three games, how should Sparky alter the pitching rotation?

9. *Transportation*—Sound Electronics, Inc. produces a battery-operated tape recorder at plants located in Martinsville, North Carolina, Plymouth, New York, and Franklin, Missouri. The unit transportation cost for shipments from the three plants to distribution centers in Chicago, Dallas, and New York are as follows:

			To	
		Chicago	Dallas	New York
	Martinsville	1.45	1.60	1.40
From	Plymouth	1.10	2.25	.60
	Franklin	1.20	1.20	1.80

After considering transportation costs, management has decided that under no circumstances will it use the Plymouth to Dallas route. In addition, a labor relations issue in the state of New York makes it impossible to ship over 100 units from Plymouth to New York during the next month. The

plant capacities and distributor orders for the next month
are as follows:

Plant	Capacity Units	Distributor	Order Units
Martinsville	400	Chicago	400
Plymouth	600	Dallas	400
Franklin	300	New York	400

Because of different wage scales at the three plants, the
unit production cost varies from plant to plant. If the costs
are $29.50/unit at Martinsville, $31.20/unit at Plymouth,
and $30.35 at Franklin, find the production and distribution
plan that minimizes production and transportation costs.

10. *Equipment acquisition*—The Two-Rivers Oil Company near
Pittsburgh transports gasoline to its distributors by trucks.
The company has recently received a contract to begin sup-
plying gasoline distributors in southern Ohio and has
$300,000 available to spend on the necessary expansion of its
fleet of gasoline tank trucks. Three types of gasoline tank
trucks are available.

Truck Model	Capacity (gals)	Purchase Cost	Monthly Operating Costs Including Depreciation
Super Tanker	5,000	$37,000	$550
Regular Line	2,500	$25,000	$425
Econo-Tanker	1,000	$16,000	$350

The company estimates that the monthly demand for the
region will be a total of 550,000 gallons of gasoline. Due
to the size and speed differences of the trucks, the different
truck models will vary in terms of the number of deliveries
or round trips possible per month. Trip capacities are estimated at 15 per month for the Super Tanker, 20 per month
for the Regular Line, and 25 per month for the Econo-
Tanker. Based on maintenance and driver availability, the
firm does not want to add more than 15 new vehicles to its
fleet. In addition, the company would like to make sure it
purchases at least three of the new Econo-Tankers to use on
the short-run, low-demand routes. As a final constraint, the

company does not want more than half of the new models to be Super Tankers.

a. If the company wishes to satisfy the gasoline demand with a minimum monthly operating expense, how many models of each truck should be purchased?

b. If the company did not require at least three Econo-Tankers and allowed as many Super Tankers as needed, what would the company strategy be?

11. *Production scheduling*—The Silver Star Bicycle Company will be manufacturing both men's and women's models for their Easy-Pedal 10-speed bicycles during the next two months, and the company would like a production schedule indicating how many bicycles of each model should be produced in each month. Current demand forecasts calls for 150 men's and 125 women's models to be shipped during the first month and 200 men's and 150 women's models to be shipped during the second month. Additional data are shown below.

Model	Produc-tion Cost	Labor Hrs. Required in Mfg. Dept.	Labor Hrs. Required in Assembly Dept.	Current Inventory
Men's	$40	10	3	20
Women's	$30	8	2	30

Last month the company used a total of 4000 man-hours of labor. The company's labor relations policy will not allow the combined total man-hours (manufacturing plus assembly) to increase or decrease by more than 500 man-hours from month to month. In addition, the company charges monthly inventory at the rate of 2% of the production cost based on the inventory levels at the end of the month. The company would like to have at least 25 units of each model in inventory at the end of the two months.

a. Establish a production schedule that minimizes production and inventory costs and satisfies the manpower smoothing, demand, and inventory requirements. What ventories will be maintained and what are the monthly man-hour requirements?

b.   If the company changed the manpower level constraints so that monthly manpower increases and decreases could not exceed 250 man-hours, what would happen to the production schedule? How much will the cost increase? What would you recommend?

12.   *Manpower balancing*—The Patriotic Doll Company manufactures two kinds of dolls: The Betsy Ross and The George Washington. The assembly process for these dolls requires two people. The assembly times are as follows:

Doll	Assembler #1	Assembler #2
Betsy Ross	6 minutes	2 minutes
George Washington	3 minutes	4 minutes
Maximum hours available per day	8 hours	8 hours

The company policy is to balance workloads on all assembly jobs. In fact, management wants to schedule work so that no assembler will have more than thirty-minutes more work per day than other assemblers. This means that in a regular eight-hour shift, all assemblers will be assigned at least 7½ hours of work. If the firm makes a $2.00 profit for each George Washington doll and $1.00 profit for each Betsy Ross doll, how many units of each doll should be produced per day? How much time will each assembler be assigned per day?

13.   *Assignment problem*—In section 5.5, we introduced the assignment problem and a new type of 0 or 1 variable. We said let $x = 1$ if an assignment is made; $x = 0$ otherwise. Remember now that linear programming does not necessarily provide integer values for the variables; however, we will now illustrate that all the extreme points of the assignment formulation happen to be integer 0 or 1 solutions. Thus, standard linear programming formulations of assignment problems will provide the necessary 0 or 1 solution values for the variables.

Consider the simple assignment problem.

Job

		1	2
	A	10	12
Person			
	B	8	9

Cost Data

a. Formulate the problem as a linear program.

b. Set up the initial tableau. Change the basis of your own choosing—any variables in solution—and observe that the solution always involves 0 and 1 valued variables.

c. List all the basic solutions for this problem.

14. *Traveling salesman*—The Lester City Steel Company has customers in Charlestown, Rossville, and Madison, Pennsylvania. The distances between the cities are summarized below:

		Lester	Charlestown	Rossville	Madison
	Lester	—	150	160	90
Miles	Charlestown	150	—	80	55
From	Rossville	160	80	—	120
	Madison	90	55	120	—

To

The regional sales manager for Lester would like to visit all three customers and return to the Lester office in such a way that he will minimize the total distance traveled. What route should the sales manager select?

Hint: Consider this as an assignment type of problem where $x_{ij} = 1$ if the sales manager travels from city i to city j and $x_{ij} = 0$ if not. We guarantee all cities are reached by making sure each city is assigned to and from one other city.

Since a city cannot be assigned to itself, the formulation of the above traveling salesmen (assignment) problem should have twelve possible assignments (variables) and eight constraints.

An additional set of constraints are required for the traveling salesman problem. If the optimal assignment involved assigning Madison to Charlestown and then assigning Charlestown back to Madison, we would not have a full loop or complete tour. The Madison-to-Charlestown-and-return loop would be unacceptable to the manager; why would he return to Madison if that is the city he has just visited? We can avoid these sub-loops or sub-tours for the above problem by adding the following constraints:

$$x_{12} + x_{21} \leq 1$$
$$x_{13} + x_{31} \leq 1$$
$$x_{14} + x_{41} \leq 1$$
$$x_{23} + x_{32} \leq 1$$
$$x_{24} + x_{42} \leq 1$$
$$x_{34} + x_{43} \leq 1.$$

As you can see, these constraints will prevent us from assigning a return route to the city the manager has just left. That is, if $x_{12} = 1$, x_{21} must be zero (unassigned).

a. Solve the above traveling salesman problem. What is the total distance traveled?

b. What happens to the solution if we don't include the sub-tour constraints? While the objective function is lower in this case, would the sales manager have a complete tour? Explain.

15. *Capital budgeting*—The Ice-Cold Refrigerator Company can invest capital funds in a variety of company projects which have different capital requirements over the next four years. Faced with limited capital resources, the company must select the most profitable projects and budget for the necessary capital expenditures. The estimated project values,

the capital requirements and the available capital projections are as follows:

Project	Estimated Present Value	Capital Requirements			
		Year 1	Year 2	Year 3	Year 4
Plant Expansion	90,000	15,000	20,000	20,000	15,000
Warehouse Expansion	40,000	10,000	15,000	20,000	5,000
New Machinery	10,000	10,000	0	0	4,000
New Product Research	37,500	15,000	10,000	10,000	10,000
Available Capital Funds		30,000	40,000	30,000	25,000

a. Which projects should the company select in order to maximize the present value of the invested funds? Show the capital budget for each year.

Hint: The decision must be made to accept or reject each project. This is similar to the assignment problem (see section 5.5) where we let $x = 1$ if the assignment is to be made and $x = 0$ if it is not to be made. Adopting a similar approach, we can let $x = 1$ if the project is accepted and $x = 0$ if it is rejected. Thus all variables must be constrained with $x \leq 1$. For this linear programming formulation of the capital budgeting problem the variables will be 0 (rejected), 1 (accepted) or a fraction between 0 and 1. Fractional values should be interpreted as insufficient funds for the complete project; therefore reject the project outright or proceed with the project in smaller increments, if possible. Actually, the method of integer programming could be used to require only 0 and 1 valued variables; however, a linear programming solution to the capital budgeting problem can still provide valuable information.

b. If the company could obtain an additional $10,000 in all four years, what would you recommend? What is the new solution?

6

Developing and Implementing Large-Scale Linear Programming Models

Prior to this chapter we have taken a step-by-step approach to studying linear programming models. In Chapters 1–4, we directed our study towards the fundamental properties of linear programming problems and the development of solution procedures for these problems. In Chapter 5, we placed a heavy emphasis on problem formulation. Although the problems in Chapter 5 were selected in order to give the reader a flavor of real-world decision-making problems, the size and scope of these problems had to be limited for pedagogical purposes.

In this chapter, however, we will investigate the overall process whereby linear programming is employed to solve large real-world problems. Specifically, we will discuss the development and implementation problems actually encountered in a large-scale linear programming project. In this context, we refer to a large-scale linear programming model of a real-world problem simply as one for which the mathematical formulation contains a very large number of decision variables and/or constraints. For the production scheduling problem we describe in this chapter, you will see that the linear programming model developed to solve the problem contains 7,200 decision variables and 6,732 constraints.

As stated then, the purpose of this chapter is to discuss both the development and implementation aspects of large-scale linear programming models. By model development we refer to the process of defining the real-world decision-making problem and then formulating a linear programming model of the problem. On the other hand, we use the term implementation to refer to the process of data preparation, computer solution, and report generation. Let us now begin our study of large-scale linear programming models by first considering the problem definition stage.

6.1   Development—Problem Definition

The problem we discuss in this chapter is an actual production scheduling problem that was encountered in a candy plant in southwestern Ohio. Ultimately, linear programming was selected as the best procedure for providing management with information about feasible, low-cost production schedules for their products. Let us see how the problem initially came to the attention of systems analyst Larry Royce. Larry was employed by the consulting firm of OR Associates located in Cincinnati, Ohio.

One day Larry was sitting in his office reviewing some computer results from his consulting project on the city bus-routing problem when he received a phone call from Bob Carter. Bob was the head of the operations analysis department for the Big Sweet Candy Company. Bob and Larry had met earlier at a monthly meeting of the local Management Science Chapter. Let us eavesdrop on their conversation.

Larry answered the phone and Bob said, "Hi, Larry. Bob Carter of Big Sweet Candy."

Larry responded, "Good to hear from you, Bob. What can I do for you?"

"Well, Larry, I think I could use your help and advice on an inventory and scheduling problem that I've been working on for the last few months. Quite frankly, I'm stumped! The problem is just too large and complex for me to solve.

"Let me go over the problem from the beginning. As you know, Big Sweet is one of the largest candy companies in the world. We manufacture over 300 products at our main plant on Howell Avenue and almost 200 products at our new plant on Fairfield Road. At these plants we use 50 to 60 types of equipment and employ approximately 350 people.

"Our problem stems from the fact that we are having difficulty determining when we should be scheduling production of our various products. Herb Flannigan, who heads our production scheduling group, has pointed out that the number of products and the number of production scheduling alternatives are too numerous to evaluate. I've talked to some of the people in Herb's department and they say they have to schedule production far in advance of when it will be needed. For example, last March we were starting to produce Christmas candy. When I checked to see why, the scheduler indicated he only had a rough idea of production capacities in later months, and that he was afraid to hold production until August or September because then he might not have enough production time to get all the candies ready for the scheduled Christmas shipments. By producing the items early, he was sure that the candies would all be available from inventory when shipment time arrived.

"I was originally given the assignment of finding procedures to cut down on our large, expensive inventory, but I have found that our schedulers hesitate to hold up production unless they are assured future machine and manpower capacities are available. Thus, they prefer to make sure we have large inventories.

"After trying to develop a procedure to keep track of machine and manpower usage and to predict machine and manpower availability for different scheduling alternatives, I have come to the conclusion that it can be done for a small number of products, but with over 500 products it's next to impossible!

"What do you think, Larry? Can you help me out?"

Obviously, Larry did not have an easy answer to the question, but he gave an encouraging reply and offered to visit Bob's office to go over the details of the situation and see what advice he could give.

Early the next Monday morning Larry arrived at Bob's office at the Howell Avenue plant. Bob took Larry on a quick tour of the production facilities, and afterwards introduced Larry to Herb Flannigan and some of the other schedulers. By talking to

Herb, Larry learned that Herb's group was using scheduling procedures that had worked quite satisfactorily in previous years when Big Sweet had a total product line of about one hundred and fifty items. But, with the addition of the new Fairfield plant and the tripling of the product line the scheduling function had become quite a headache.

Larry and Bob returned to the office and over coffee Larry offered some suggestions.

Larry said, "Bob, your problem is not going to be easy to solve, but my thinking is that we should work toward developing a total computer package that will provide the information you want on a routine basis. Hopefully, we can design a system that will require little or no modification as your product line and production facilities change in the future; that is, the system must be flexible to meet your changing needs.

"I think we can best approach this problem by considering computerized quantitative procedures such as dynamic programming, linear programming, or a specialized scheduling model. I'll spend some time evaluating these alternatives. What we need to do today is to outline as clearly as possible the objectives you want met by a computerized scheduling procedure and what conditions or constraints must be satisfied in order to have a workable production schedule." The rest of the day was spent discussing this issue. By late afternoon Bob and Larry had agreed on a verbal statement of the objectives and the constraints. Their notes are shown in Table 6–A.

Larry and Bob agreed that the output of the computerized scheduling procedure should be a production plan that specified how many units of each product should be scheduled for production in each of the next twelve months. This scheduling procedure would guarantee that future machine and manpower capacities would be available to meet future production requirements. For example, if the schedule called for the production of some of the Christmas candy in October, the scheduler could be assured that capacities would be available and that an earlier production of these products would be unnecessary. This twelve-month production plan could be revised or updated as frequently as desired, but Larry and Bob both agreed that a monthly revision should be prepared on a routine basis.

Objectives: A routine computer procedure that develops a production plan having the following properties:

1. selects minimum cost production alternatives if two or more machines can produce the same item (i.e., minimizes production costs)

2. provides for minimum inventory levels (i.e., minimizes inventory cost)

Constraints:

1. Production capacity—the schedule cannot require more machine time than is available

2. Manpower capacity—the schedule cannot require more labor time than is available

3. Demand—all items must be ready by the shipment date

4. Safety stock levels—a buffer stock should be available in case demand exceeds expectations

5. Shelf life conditions—maximum time products can stay in inventory without spoiling

6. Storage space limitations—warehouse capacities

7. Manpower smoothing—workforce levels should not fluctuate excessively

Table 6–A.   Objectives and Constraints for the Big Sweet Candy Company Scheduling Procedure

Let us go back and see what we have done thus far. Actually we are working on the first step of the development process: problem definition. Larry has done a good job of helping Bob pin-point some of the details of his problem. Up until now, Bob had been overwhelmed by what he termed a problem "too hard and complex for me to solve". But after meeting with Larry, Bob at least had a clear definition of the problem with respect to his real objectives and his real constraints.

Keep in mind that while the problem definition step is the first phase of virtually all decision-making problems, it is not a trivial step. Usually the problem starts with a rather broad or general description that takes time, imagination, and effort to transform into a well-defined problem. Note that our scheduling project started as an "excessive inventory problem" and was eventually defined more concisely in terms of the two-part objective and seven-constraint problem described in Table 6–A.

In addition, you should be aware of the fact that it is almost certain that the first problem definition will be revised several times before the problem is eventually solved. Usually the revisions are based on additional constraints and/or a revised objective that becomes known to the analyst as he works toward solving the problem. If we have done a thorough job with our first problem definition, these later revisions will usually be minor.

After defining the problem, we must give some thought as to the best way to solve it. We must not restrict ourselves to one or two solution procedures; but rather, we must be open-minded and prepared to consider a variety of procedures. This was the approach that Larry took. By discussing the problem with some of his colleagues back at the office, Larry was able to discard several preliminary alternatives and finally arrive at a conclusion that a linear programming model offered good potential for a successful solution.

Once it was decided that a linear programming model would be used, Larry was ready to proceed to the next step in the development process: problem formulation.

6.2   Development—Problem Formulation

We will now present a detailed formulation of the linear programming model which was developed for scheduling the candy plant's products. As you will soon learn, the notation required by a large-scale linear programming problem becomes somewhat more sophisticated than the notation we used in the production scheduling application described in Section 5.3. Specifically, you will now see decision variables with three subscripts instead of the two subscripted decision variables you have been exposed to in Chapter 5. We recommend that you review the two-product, three-month production scheduling application of Section 5.3 before attempting to study this section. Initially, you should just read this section for the general concepts. Then, on the second reading, you can pick up the details of the notation and problem formulation. As you will see, the linear programming model described in this section presents a complete package that might be used to solve a variety of similar scheduling problems.

As stated above, triple subscript notation is used to define the decision variables for our production scheduling problem. The first subscript is used to indicate the product, the second subscript to indicate the route, and the third subscript to indicate the month. Our decision variables can then be represented as

x_{irm} = units of product i produced via route r in month m.

To illustrate the use of this triple subscript notation, consider the following illustration. Suppose that Bob's original problem had involved the production of two products (A and B) over a two-month planning period. Let us assume that product A can be produced on either of two production lines which we will identify as production routes 1 and 2 for product A. In addition, let us assume that product B can be produced on only one machine and therefore has only one possible production route. If we ask you to define the decision variables for this two-product, two-month scheduling problem in terms of triple subscripted decision variables, x_{irm}, you should come up with something like the following:

x_{111} = units of product A produced on route 1 in month 1
x_{112} = units of product A produced on route 1 in month 2
x_{121} = units of product A produced on route 2 in month 1
x_{122} = units of product A produced on route 2 in month 2
x_{211} = units of product B produced on route 1 in month 1
x_{212} = units of product B produced on route 1 in month 2.

Familiarity with this notation will be a great aid both in the formulation of the problem and in the interpretation of the output. Suppose that the linear programming solution to the above two-product, two-month problem provided the following result:

$$x_{212} = 700.$$

If you understand our triple subscript notation, you can see that this production schedule calls for 700 units of product B to be produced on route 1 in month 2.

Let us now return to our two-product, two-month problem and find the total number of units of both products produced over the two-month period. The total production (TP) is simply the sum of all production variables.

$$TP = x_{111} + x_{112} + x_{121} + x_{122} + x_{211} + x_{212}$$

Using summation notation $\sum$, we can write the above expression in a more concise form.

$$TP = \sum_{i=1}^{2} \sum_{r=1}^{R_i} \sum_{m=1}^{2} x_{irm}$$

$$= x_{111} + x_{112} + x_{121} + x_{122} + x_{211} + x_{212}$$

where R_i = the number of different production routes available for product i

Finally, if we let

N = number of products, and

M = the number of months,

then we can adopt the following more general expression for total production (TP):

$$TP = \sum_{i=1}^{N} \sum_{r=1}^{R_i} \sum_{m=1}^{M} x_{irm}. \tag{6.1}$$

The above notation can be considered a general form because it provides the total production if we have $N = 2$ products or $N = 500$ products, if we have $R_i = 1$ route or $R_i = 3$ routes, and if we have $M = 2$ months or $M = 12$ months.

In the Big Sweet Candy problem production schedules will have to be generated for 500 products ($N = 500$). Since 100 of these products have a second route alternative, we have a total of 600 product-route combinations. Thus, if management wants the linear programming solution to determine a production quantity for each product-route combination in each of 12 months ($M = 12$), we will need a total of 7200 = 600 x 12 decision variables (x_{irm}). For this problem the summation given in equation (6.1) will have 7200 terms. Hopefully, the use of the summation sign will not cause the reader any additional difficulty in interpreting the constraints and the objective function for our large-scale linear program. The only reason for its use is to make the notation more concise. Without the use of a summation sign, it would take an entire chapter of this book to write out a single constraint for our problem.

Let us now proceed to a mathematical formulation of the constraints for our problem.

Production Capacity Constraints

The candy plant operates with over fifty different pieces of production equipment. After allowing for routine maintenance and repair time, each production department can notify the schedulers as to how many hours each piece of equipment will be available each month. The schedulers must then develop a production plan that will not require more total production time than is available on each machine.

For example, let us suppose the production department reports that machine number three has 150 hours of production time available for the first month. Since x_{ir1} indicates the number of units of product i to be produced on route r in month 1, we can determine the total time that machine three is being used for production if we know the time required for each unit. We let

C_{em} = production hours available for equipment e in month m, and

t_{ire} = production time, in hours, required to produce product i via route r on equipment e.

If, as stated about, 150 hours of production time is available on machine three in month one, then $C_{31} = 150$. If we let $t_{ire} = 0$ when product i produced via route r does not use equipment e, then the total production time used by all products on machine three in month one is given by

$$\sum_{i=1}^{N} \sum_{r=1}^{R_i} t_{ir3} x_{ir1}.$$

The above expression simply sums the production times required for all products using machine three in month one. The particular production capacity constraint for this machine in month one becomes

$$\sum_{i=1}^{N} \sum_{r=1}^{R_i} t_{ir3} x_{ir1} \leq 150.$$

We must have one of these constraints for every piece of equipment in every period.

The general expression for the production capacity constraints is given by

$$\sum_{i=1}^{N} \sum_{r=1}^{R_i} t_{ire} x_{irm} \leq C_{em} \qquad (6.2)$$

$$\text{for} \quad e = 1, 2, \ldots, E$$
$$m = 1, 2, \ldots, M.$$

In this expression E is used to represent the total number of pieces of equipment. Since, for our candy plant problem, we have $E = 50$ machines and $M = 12$ months, we will have a total of $50 \times 12 = 600$ production capacity constraints of the form of inequality (6.2).

Manpower Capacity Constraints

The personnel department projects the total man-hours of production time available each month for each of three labor grades: skilled labor, unskilled labor-class A, and unskilled labor-class B. Fluctuations in the labor hours available each month are caused by vacations and some standard seasonal layoff patterns. The scheduler's job is now complicated by the fact that he must guarantee that the production plan will not require more man-hours than are available.

Our manpower capacity constraints are similar in form to the production capacity constraints. Let

L_{gm} = man-hours available for labor grade g in month m,
G = number of labor grades, and

h_{irg} = man-hours of labor grade g required to produce one
          unit of product i via route r.

The total manpower requirements for labor grade g in month m are then given by

$$\sum_{i=1}^{N} \sum_{r=1}^{R_i} h_{irg}x_{irm}.$$

Our manpower capacity constraints can then be written as:

$$\sum_{i=1}^{N} \sum_{r=1}^{R_i} h_{irg}x_{irm} \leq L_{gm} \qquad (6.3)$$

$$\text{for} \quad g = 1, 2, \ldots, G$$
$$m = 1, 2, \ldots, M.$$

Since we have $G = 3$ labor grades and $M = 12$ months, we have $3 \times 12 = 36$ manpower capacity constraints of the above form.

Demand Constraints

Each month the company's forecasting group issues a one-year demand projection which indicates how many units of each product must be shipped each month. We now have to develop the linear programming constraints that guarantee all products will be available by the shipment deadlines.

In making shipments for any one month we can use items that have been produced in the current month or items that have been produced in previous months and are now available from inventory. Hence, to meet the monthly demand for each product we must require that the production volume for the month plus the inventory remaining from production in previous months exceeds the amount to be shipped.

That is,

Current Production + Current Inventory ≥ Demand.

Introducing the notation

d_{im} = forecasted demand in units of product i to be shipped in month m, and

s_{im} = inventory in units of product i on hand at the end of month m, we can see that the current inventory for month m is actually the inventory that existed at the end of the previous month $[s_{i,(m-1)}]$; thus, to satisfy the demand for any one product in any given month, we must require that

$$\sum_{r=1}^{R_i} x_{irm} + s_{i,(m-1)} \geq d_{im}. \tag{6.4}$$

The above constraint guarantees that the total production of product i in month m plus the inventory carried from the previous month, (m–1), will be sufficient to satisfy the demand in month m.

Inequality (6.4) is correct, but it leaves the schedulers with one major problem: they cannot determine the inventory at the end of period m–1 because they are not as yet aware of how many units of the product have been scheduled and produced in the previous m–1 months. That is, $s_{i,(m-1)}$ is dependent upon the production variables $x_{ir1}, \ldots, x_{ir(m-1)}$, and we do not know these values until we have solved the linear programming problem. Therefore, we must look for ways to replace the ending inventory term, $s_{i,(m-1)}$, in the above constraints. Our approach will be to find a way of expressing the ending inventory in terms of the production variables, the known demand values, and the known inventory level that existed at the start of the production schedule. This starting inventory level is frequently referred to as the beginning inventory. Using these quantities we find that

$$\begin{array}{c} \text{Ending} \\ \text{Inventory} \\ \text{for period m} \end{array} = \begin{array}{c} \text{Beginning} \\ \text{Inventory} \end{array} + \begin{array}{c} \text{Total Production} \\ \text{for m months} \end{array} - \begin{array}{c} \text{Total Shipments} \\ \text{for m months.} \end{array}$$

Letting s_{i0} indicate the beginning inventory for product i we have

$$s_{im} = s_{i0} + \sum_{j=1}^{m} \sum_{r=1}^{R_i} x_{irj} - \sum_{j=1}^{m} d_{ij}. \tag{6.5}$$

We have used the subscript j in the above equation to indicate the month in which the productions and demands occurred. Subscript j serves the same purpose as subscript m in the previous constraints.

Using equation (6.5), we can similarly write an expression for the inventory level at the end of month m–1.

$$s_{i(m-1)} = s_{i0} + \sum_{j=1}^{m-1} \sum_{r=1}^{R_j} x_{irj} - \sum_{j=1}^{m-1} d_{ij} \qquad (6.6)$$

Using the above equation for $s_{i(m-1)}$ in our original demand constraint (6.4), we have

$$\sum_{r=1}^{R_i} x_{irm} + s_{i0} + \sum_{j=1}^{m-1} \sum_{r=1}^{R_j} x_{irj} - \sum_{j=1}^{m-1} d_{ij} \geq d_{im}. \qquad (6.7)$$

Recall that our standard form for linear programming constraints requires us to put the variables on the left-hand side of the inequality and the known constants on the right-hand side. Since the production variables (x_{irj}) are the unknowns, and the demands (d_{ij}) and beginning inventory (s_{i0}) are the known constants, we can rearrange inequality (6.7) to obtain the following form for our demand constraints.

$$\sum_{r=1}^{R_i} x_{irm} + \sum_{j=1}^{m-1} \sum_{r=1}^{R_j} x_{irj} \geq d_{im} + \sum_{j=1}^{m-1} d_{ij} - s_{i0} \qquad (6.8)$$

Since

$$\sum_{r=1}^{R_i} x_{irm} + \sum_{j=1}^{m-1} \sum_{r=1}^{R_j} x_{irj} = \sum_{j=1}^{m} \sum_{r=1}^{R_j} x_{irj}, \qquad (6.9)$$

and

$$d_{im} + \sum_{j=1}^{m-1} d_{ij} = \sum_{j=1}^{m} d_{ij}, \qquad (6.10)$$

we can write our demand constraints as follows:

$$\sum_{j=1}^{m} \sum_{r=1}^{R_i} x_{irj} \geq \sum_{j=1}^{m} d_{ij} - s_{i0}. \qquad (6.11)$$

We must have one of these constraints for all products ($i = 1, 2, \ldots, N$) and all months ($m = 1, 2, \ldots, M$). Hence, in our problem we need $(500) \times (12) = 6{,}000$ additional constraints to guarantee that demand will be met.

Safety Stock Constraints

Safety stock levels indicate the number of units of each product that must be maintained in inventory in excess of what is needed to satisfy forecasted demand. The purpose of the safety stock is to meet unexpected emergencies such as actual demands exceeding the forecasted volumes and possible production equipment breakdowns that would cause higher than expected shipments from inventory. The safety stock levels for the products were stated such that ending inventories for each product in each month had to be sufficient to cover at least one-half of the shipment volume scheduled for the next month. This in effect gave the company approximately a two-week safety period to adjust production schedules to meet emergency conditions. An advantage of this method of defining safety stock is that during the low demand seasons a one-half month safety stock will be very small, while during the high demand months which usually exhibit greater uncertainties high safety stock levels will be maintained.

Given the requirements for safety stock which have been outlined above we see that the safety stock constraints for our candy plant problem may be expressed very simply as

$$s_{im} \geq \tfrac{1}{2} d_{i(m+1)} \tag{6.12}$$

where $d_{i(m+1)}$ is the next month's projected demand.

Using equation (6.5) as our means of expressing ending inventory, we have

$$s_{i0} + \sum_{j=1}^{m} \sum_{r=1}^{R_i} x_{irj} - \sum_{j=1}^{m} d_{ij} \geq \tfrac{1}{2} d_{i(m+1)}. \tag{6.13}$$

Again working to place all unknown variables on the left-hand side and the constant terms on the right-hand side, we may rewrite our complete set of safety stock constraints.

$$\sum_{j=1}^{m} \sum_{r=1}^{R_i} x_{irj} \geq \sum_{j=1}^{m} d_{ij} + \tfrac{1}{2} d_{i(m+1)} - s_{i0} \qquad (6.14)$$

$$\text{for} \quad i = 1, 2, \ldots N$$
$$m = 1, 2, \ldots M$$

Again we have a total number of additional constraints of $(500) \times (12) = 6,000$.

Our large-scale problem is continuing to grow in size; however, if you have been watching closely you may have observed a way in which we can right now begin to reduce the number of constraints. Specifically, compare our demand constraints (6.11) and our safety stock constraints (6.14). As you can see they are identical except that the right-hand side of the safety-stock constraints contain the additional term $\tfrac{1}{2} d_{i(m+1)}$. Since the demand variables are never negative, the right-hand side of the safety-stock constraints must be greater than or equal to the right-hand side of the demand constraints. If you recall the redundancy concept defined in Chapter 1, you know that a redundant constraint is one that is unnecessary, since if all of the other constraints are satisfied it will automatically be satisfied. Hence, any redundant constraints can be dropped from the problem without having any effect on the optimal solution. Are there any redundant constraints in our problem? Since the right-hand side of equation (6.14) is always larger than the right-hand side of equation (6.11), satisfying the safety stock constraints (6.14) always guarantees that the demand constraints will be satisfied. Thus, the 6,000 demand constraints are redundant and can be dropped from the formulation.

Intuitively, we can see that if we have produced enough to meet the safety stock conditions, we must have satisfied the demand for that month because the safety stock is actually the inventory left at the end of a month after all shipments have been made.

Redundant equations are not always so readily identified; however, in formulating any linear programming problem you should make every effort to identify such constraints and omit them from the formulation. This is especially true in a large-scale system where a reduction of a few thousand constraints may result in significant savings in terms of data preparation and computer time.

Shelf-Life Constraints

A significant part of the inventory problem in the food product industry is the limitation on the length of time items may remain in inventory. In Big Sweet candy plants the schedulers had been producing some Christmas products in March and April. These items remained in inventory for periods of time ranging from six to eight months. Based on considerations of quality and freshness this was undesirable. Ideally food products should have a relatively low inventory or shelf life or in other words, a relatively high inventory turnover rate.

Specifically in the candy plant problem we are considering, the Quality Control Department specified a maximum number of months that each product could remain in inventory. Suppose that the current inventory level for one particular product is 100 units and that the Quality Control Department has placed a three-month life on this product. If we have a total demand during the next three months of 100 or more units, a first-in-first-out inventory policy will guarantee that all 100 units will be shipped within the specified shelf-life time limit. If, however, the demand during the three-month shelf-life period is less than 100 units, we know at least some of our current inventory will still be in inventory four months from now, and thus the inventory shelf-life restrictions will be violated.

We can incorporate inventory shelf-life constraints into our linear programming formulation by making sure that inventory levels at the end of each month are always less than or equal to the total demand for the products over their shelf-life period. The shelf-life constraints are shown below.

Let

w_i = inventory life expressed in the number of months product i may remain in the warehouse (w_i was always at least one month)

If we are in month m, the demand during the inventory life of w_i months for product i is given by

$$\sum_{j=m+1}^{m+w_i} d_{ij}.$$

You should note that we are assuming that the inventory life, w_i, will always be an integer. Thus our inventory life constraints can be written as

$$s_{im} \leq \sum_{j=m+1}^{m+w_i} d_{ij}. \tag{6.15}$$

Again using the ending inventory expression (6.5) we can rewrite (6.15) as

$$s_{i0} + \sum_{j=1}^{m} \sum_{r=1}^{R_i} x_{irj} - \sum_{j=1}^{m} d_{ij} \leq \sum_{j=m+1}^{m+w_i} d_{ij}.$$

Finally, placing only the unknown variables on the left-hand side of the constraints, we have the following set of inventory life constraints:

$$\sum_{j=1}^{m} \sum_{r=1}^{R_i} x_{irj} \leq \sum_{j=1}^{m+w_i} d_{ij} - s_{i0} \tag{6.16}$$

$$\text{for} \quad i = 1, 2, \ldots, N$$
$$m = 1, 2, \ldots, M.$$

We have again added $(500) \times (12) = 6,000$ constraints to our problem.

Storage-Space Constraints

Storage-space constraints are needed to make sure inventory levels do not exceed the storage-space available in each type of warehouse storage. In our production scheduling problem, two types of storage areas were defined: conditioned storage and general storage. The conditioned storage had special temperature and humidity control and was used to store products that would spoil without a controlled environment. In the general storage area no attempt was made to provide a controlled environment. Candy products stored in this area were basically insensitive to normal temperature and humidity fluctuations.

These storage-space constraints were established by limiting the size of the ending inventories. Let

H_{km} = the warehouse capacity in cubic feet of storage type k available in month m,

k = the type of storage area (k = 1 for conditioned storage and k = 2 for general storage),

K = the number of different storage types (2 in our problem), and

r_{ik} = the number of cubic feet of storage type k required by one unit of product i (r_{ik} = 0 if product i does not utilize storage type k).

The storage-space constraints for all products stored in area k during period m, can now be written as

$$\sum_{i=1}^{N} r_{ik}s_{im} \leq H_{km}. \tag{6.17}$$

Here we see that $r_{ik}s_{im}$ is the number of cubic feet of storage space type k used by product i in month m. Hence, we have simply summed all products to get the storage-space constraint, (6.17). Using equation (6.5) to eliminate s_{im} from inequality (6.17), we can write our storage-space constraints in the following form:

$$\sum_{i=1}^{N} r_{ik} \left[s_{io} + \sum_{j=1}^{m} \sum_{r=1}^{R_i} x_{irj} - \sum_{j=1}^{m} d_{ij} \right] \leq H_{km}.$$

Rearranging terms, we get

$$\sum_{i=1}^{N} r_{ik} \sum_{j=1}^{m} \sum_{r=1}^{R_i} x_{irj} \leq H_{km} + \sum_{i=1}^{N} r_{ik} \sum_{j=1}^{m} d_{ij} \sum_{i=1}^{N} r_{ik}s_{io} \tag{6.18}$$

$$\text{for} \quad k = 1, \ldots, K$$
$$m = 1, \ldots, M.$$

Note that the right-hand side of the above constraints contains known values for H_{km}, r_{ik}, d_{ij}, and s_{io}. When we specify these values, the right-hand side of each constraint will become one numerical value, and the storage-space constraints will be in the desired form.

We will require one of the above constraints for each type of storage in each month. This gives us a total of $2 \times 12 = 24$ storage-space constraints.

Manpower Smoothing Constraints

To gain an understanding of what manpower smoothing constraints attempt to do, consider the following illustration. Suppose that in one month the plant operated with 200 people. Clearly, the company would experience some problems if it selected a production schedule that required an increase in manpower to 400 people for the next month. Why? The major problem with such a big manpower change is that it may be difficult to find, hire, and train personnel necessary to reach a labor force of 400 people in a one-month period. Similarly, a sudden reduction in the labor force could have an undesirable effect on employee morale and might cause union problems. The purpose of manpower smoothing constraints is to keep the optimal solution to our linear programming model from calling for a wide variety of ups and downs in our total labor force requirements.

In the production scheduling problem illustrated in Section 5.3 we presented one approach for handling monthly variations in the total labor force. As you recall, the approach presented involved defining decision variables that reflected the increase or decrease in the number of man-hours used from one month to another. By assigning appropriate cost coefficients for these variables and including these variables as part of the objective function, we saw how large ups and downs in our labor force could be prevented. In this section we will present an alternative method for limiting production fluctuations. The method will involve formulating constraints that will place limits on the amount of fluctuation that can occur from one month to another. In contrast to the approach illustrated in Chapter 5, this new method will not require defining any new decision variables.

To see how labor force fluctuations are handled using this new approach, let us first recall some of our previous notation.

x_{irm} = units of product i produced via route r in month m

h_{irg} = man-hours of labor grade g required to produce one unit of product i via route r

The total man-hours of labor grade g used in month m is given by

$$\sum_{i=1}^{N} \sum_{r=1}^{R_i} h_{irg} x_{irm}. \tag{6.19}$$

How many employees will be required in order to provide the above total man-hours? This can be calculated as follows:

Let z_m = the number of man-hours worked by one employee in month m.

z_m is determined by multiplying the number of working days in month m by the number of hours normally worked in a regular shift. For example, a twenty work-day month with standard eight-hour shifts would result in a z_m of 20 x 8 = 160. Therefore, the total number of employees of labor grade g needed in period m is then

$$\frac{1}{z_m} \sum_{i=1}^{N} \sum_{r=1}^{R_i} h_{irg} x_{irm}. \tag{6.20}$$

Suppose we let

A_g = actual number of employees of labor grade g used in the month prior to the start of our scheduling period.

Since A_g is historical manpower usage, the schedulers will know its value. Now what is the change in the total number of employees of labor grade g required to meet the production scheduled in the first month? This is given as follows:

$$\text{Net Manpower Change for Month 1} = \text{Manpower Needs Month 1} - \text{Actual Manpower Used in Previous Month}$$

or

$$\text{Net Change} = \frac{1}{z_1} \sum_{i=1}^{N} \sum_{r=1}^{R_i} h_{irg} x_{ir1} - A_g. \tag{6.21}$$

If this net change is zero, we have done a perfect job of smoothing manpower requirements. That is, the number of employees needed in the first month will be the same as the number of employees working the previous month (A_g). A positive net change means that the company must increase its labor force, while a negative net change means that the company must decrease its labor force.

We can control labor fluctuations by requiring in our linear programming formulation that the net change between months not exceed maximum reasonable values. For example, assume that for the first month management wanted to restrict labor fluctuations to a maximum increase of 50 employees and a maximum decrease of 25 employees. The following labor constraints will provide for these maximum manpower changes.

$$\text{Net Change} \leq 50 \quad \text{Maximum Increase}$$
$$\text{Net Change} \geq -25 \quad \text{Maximum Decrease}$$

or

$$\frac{1}{z_1} \sum_{i=1}^{N} \sum_{r=1}^{R_i} h_{irg}x_{ir1} - A_g \leq 50$$

$$\frac{1}{z_1} \sum_{i=1}^{N} \sum_{r=1}^{R_i} h_{irg}x_{ir1} - A_g \geq -25$$

To get the right-hand side of our maximum decrease constraint positive, we multiply each side of this constraint by -1. Thus, we get

$$A_g - \frac{1}{z_1} \sum_{i=1}^{N} \sum_{r=1}^{R_i} h_{irg}x_{ir1} \leq 25.$$

In general, let

U_{gm} = maximum allowable manpower increase for labor grade g in month m, and

V_{gm} = maximum allowable manpower decrease for labor grade g in month m,

for $g = 1, 2, \ldots, G$
     $m = 1, 2, \ldots, M.$

(Note that U_{gm} and V_{gm} are known constants and input data to the linear programming model.)   Our constraints for month 1 can now be written as

$$\frac{1}{z_1} \sum_{i=1}^{N} \sum_{r=1}^{R_i} h_{irg} x_{ir1} - A_g \leq U_{g1}$$

$$A_g - \frac{1}{z_1} \sum_{i=1}^{N} \sum_{r=1}^{R_i} h_{irg} x_{ir1} \leq V_{g1}.$$

The manpower smoothing constraints that we have established thus far take into account only the labor changes for month one. Let us look at the smoothing constraints for the other months. In particular, consider month two.   The actual number of employees of labor grade g required during this month is given by equation (6.20) with m = 2.   As we have already seen, the labor requirement for month one is given by the same equation with m = 1.   Therefore, the net change in the number of employees of labor grade g between months 1 and 2 is given by:

$$\text{Net Change} = \frac{1}{z_2} \sum_{i=1}^{N} \sum_{r=1}^{R_i} h_{irg} x_{ir2} - \frac{1}{z_1} \sum_{i=1}^{N} \sum_{r=1}^{R_i} h_{irg} x_{ir1}.$$

Thus, we must add the following manpower smoothing constraints for labor grade g in month 2.

$$\frac{1}{z_2} \sum_{i=1}^{N} \sum_{r=1}^{R_i} h_{irg} x_{ir2} - \frac{1}{z_1} \sum_{i=1}^{N} \sum_{r=1}^{R_i} h_{irg} x_{ir1} \leq U_{g2}$$

$$\frac{1}{z_1} \sum_{i=1}^{N} \sum_{r=1}^{R_i} h_{irg} x_{ir1} - \frac{1}{z_2} \sum_{i=1}^{N} \sum_{r=1}^{R_i} h_{irg} x_{ir2} \leq V_{g2}.$$

The above two constraints for month two are similar to the constraints required for all subsequent months. The general form for these constraints can be written as follows:

$$\frac{1}{z_m} \sum_{i=1}^{N} \sum_{r=1}^{R_i} h_{irg}x_{irm} - \frac{1}{z_{(m-1)}} \sum_{i=1}^{N} \sum_{r=1}^{R_i} h_{irg}x_{ir(m-1)} \leq U_{gm} \quad (6.22)$$

$$\frac{1}{z_{(m-1)}} \sum_{i=1}^{N} \sum_{r=1}^{R_i} h_{irg}x_{ir(m-1)} - \frac{1}{z_{m}} \sum_{i=1}^{N} \sum_{r=1}^{R_i} h_{irg}x_{irm} \leq V_{gm} \quad (6.23)$$

$$\text{for} \quad g = 1, 2, \ldots, G$$
$$m = 1, 2, \ldots, M.$$

In total we will have two manpower smoothing constraints for each labor grade in each month. This gives us $(2) \times (3) \times (12) = 72$ additional constraints.

Objective Function

Now the question is what do we optimize? That is, what should our objective function be? Bob and Larry's original problem definition (Table 6–A) included two objectives:

1. minimize production costs
2. minimize inventory costs.

Our goal now is to establish a total cost function for our linear programming model. We will then of course try to find the production schedule that satisfies all our constraint conditions at a minimum total cost.

Let us consider the cost components one at a time. The production costs refer to the labor and material costs associated with producing one unit of product. Since several of the products have more than one production route alternative, the cost per unit will vary with the route selected. We cannot always use the minimum cost route for all products because many times we will find the minimum cost routes for two similar products require the same machine. If the products must be produced at the same point in time, one of the products will have to be assigned an alternative route. The alternative route selected will have to be based on consideration of the availability of the alternative route equipment and on the cost of other products that might be produced on the alternative route equipment. Thus, we will want the linear programming model to identify the production route alternatives for each product.

Let c_{ir} = total labor and material costs for one unit of product i produced via route r.

The portion of the objective function representing production costs (PC) may be written as

$$PC = \sum_{i=1}^{N} \sum_{r=1}^{R_i} \sum_{m=1}^{M} c_{ir} x_{irm}. \tag{6.24}$$

For our candy plant problem it turns out that the materials cost is the same regardless of the production route; therefore c_{ir} can be defined solely in terms of the labor production cost per unit.

The inventory cost is based on the size of the inventories and the cost of carrying a unit of the product in inventory for a given period of time.

Let q_i = cost of carrying one unit of product i in inventory for one month,
$\quad$ IC = inventory cost
$\quad s_{im}$ = units of product i in inventory at the end of month m.

Using this notation we see that

$$IC = \sum_{i=1}^{N} \sum_{m=1}^{M} q_i s_{im}. \tag{6.25}$$

The ending inventory level (s_{im}) is used as the measure of inventory size. As we noted in equation (6.5), the ending inventory can be expressed in terms of the beginning inventory, production quantities, and demand. Using equation (6.5), we may rewrite equation (6.25) as follows:

$$IC = \sum_{i=1}^{N} \sum_{m=1}^{M} q_i \left(s_{i0} + \sum_{j=1}^{m} \sum_{r=1}^{R_i} x_{irj} - \sum_{j=1}^{m} d_{ij} \right)$$

$$= \sum_{i=1}^{N} \sum_{j=1}^{M} q_i s_{i0} + \sum_{i=1}^{N} \sum_{m=1}^{M} q_i \sum_{j=1}^{m} \sum_{r=1}^{R_i} x_{irj}$$

$$\sum_{i=1}^{N} \sum_{m=1}^{M} q_i \sum_{j=1}^{m} d_{ij}. \tag{6.26}$$

The first and third terms of the inventory cost equation, (6.26), are constant regardless of the production schedule. That is, no matter which production schedule we choose beginning inventories and demand will not be affected. Therefore, it is not necessary to include these terms of the objective function. The second term contains the portion of inventory cost that varies with the production schedule. Thus, we can write the variable portion of our inventory cost (denoted IC') as follows:

$$IC' = \sum_{i=1}^{N} \sum_{m=1}^{M} \sum_{j=1}^{m} \sum_{r=1}^{R_i} q_i x_{irj}. \tag{6.27}$$

Let us take a particular product i and route r and see what happens to the above equation as we consider the inventory cost over three months. That is, let m vary from 1 to 3. For this particular case, equation (6.27) becomes

$$\sum_{m=1}^{3} \sum_{j=1}^{m} q_i x_{irj} = q_i x_{ir1} + q_i x_{ir1} + q_i x_{ir2}$$

$$+ q_i x_{ir1} + q_i x_{ir2} + q_i x_{ir3}$$

$$= 3q_i x_{ir1} + 2q_i x_{ir2} + q_i x_{ir3}. \tag{6.28}$$

This pattern would continue if we considered more than three months. Note that equation (6.28) could be written as

$$\sum_{m=1}^{M} (M - m + 1) q_i x_{irm}. \tag{6.29}$$

Thus we can substitute expression (6.29) into the variable inventory carrying cost equation (6.27).

$$IC' = \sum_{i=1}^{N} \sum_{r=1}^{R_i} \sum_{m=1}^{M} (M - m + 1) q_i x_{irm} \qquad (6.30)$$

Note that the production cost (equation (6.24)) and the variable inventory cost (equation (6.30)) are both functions of the production variable x_{irm}. Combining these two costs in one equation, we have

$$TC = \text{Total Cost} = PC + IC'$$

$$= \sum_{i=1}^{N} \sum_{r=1}^{R_i} \sum_{m=1}^{M} [c_{ir} + (M - m + 1) q_i] x_{irm}. \qquad (6.31)$$

The coefficient $[c_{ir} + (M - m + 1) q_i]$ must be specified for each production variable, x_{irm}. The values are the cost coefficients included in the objective function.

The Complete Model

Collecting together all of the constraints and the objective function we may now present the complete mathematical formulation of the candy plant's large-scale linear programming model.

Objective Function:

$$\min \sum_{i=1}^{N} \sum_{r=1}^{R_i} \sum_{m=1}^{M} [c_{ir} + (M - m + 1) q_i] x_{irm} \qquad (6.31)$$

s.t.

Production Capacity Constraints:

$$\sum_{i=1}^{N} \sum_{r=1}^{R_i} t_{ire} x_{irm} \leq C_{em} \quad \begin{matrix} e = 1, \ldots, E \\ m = 1, \ldots, M \end{matrix} \qquad (6.2)$$

Manpower Capacity Constraints:

$$\sum_{i=1}^{N} \sum_{r=1}^{R_i} h_{irg} x_{irm} \leq L_{gm} \qquad \begin{array}{l} g = 1, \ldots, G \\ m = 1, \ldots, M \end{array} \qquad (6.3)$$

Safety Stock Constraints:

$$\sum_{j=1}^{m} \sum_{r=1}^{R_i} x_{irj} \geq \sum_{j=1}^{m} d_{ij} + \tfrac{1}{2} d_{i(m+1)} - s_{i0} \qquad \begin{array}{l} i = 1, \ldots, N \\ \\ m = 1, \ldots, M \end{array} \qquad (6.14)$$

Shelf-Life Constraints:

$$\sum_{j=1}^{m} \sum_{r=1}^{R_i} x_{irj} \leq \sum_{j=1}^{m+w_i} d_{ij} - s_{i0} \qquad \begin{array}{l} i = 1, \ldots, N \\ \\ m = 1, \ldots, M \end{array} \qquad (6.16)$$

Storage-Space Constraints:

$$\sum_{i=1}^{N} r_{ik} \sum_{j=1}^{m} \sum_{r=1}^{R_i} x_{irj} \leq H_{km} + \sum_{i=1}^{N} r_{ik} \sum_{j=1}^{m} d_{ij} - \sum_{i=1}^{N} r_{ik} s_{i0} \qquad (6.18)$$

$$k = 1, \ldots, K$$
$$m = 1, \ldots, M$$

Manpower Smoothing Constraints:

$$\frac{1}{z_m} \sum_{i=1}^{N} \sum_{r=1}^{R_i} h_{irg} x_{irm} - \frac{1}{z_{(m-1)}} \sum_{i=1}^{N} \sum_{r=1}^{R_i} h_{irg} x_{ir(m-1)} \leq U_{gm} \qquad (6.22)$$

$$g = 1, \ldots, G$$
$$m = 1, \ldots, M$$

$$\frac{1}{z_{(m-1)}} \sum_{i=1}^{N} \sum_{r=1}^{R_i} h_{irg} x_{ir(m-1)} - \frac{1}{z_m} \sum_{i=1}^{N} \sum_{r=1}^{R_i} h_{irg} x_{irm} \leq V_{gm} \qquad (6.23)$$

$$g = 1, \ldots, G$$
$$m = 1, \ldots, M$$

$$i = 1, \ldots, N$$
$$x_{irm} \geq 0 \quad r = 1, \ldots, R_i$$
$$m = 1, \ldots, M$$

Problem Size

The number of variables and constraints for our linear programming formulation are summarized in Table 6–B.

Variables

Production variable (x_{irm})	7,200

Constraints

1.	Production capacity	600
2.	Manpower capacity	36
3.	Demand	----
4.	Safety stock	6,000
5.	Shelf life	6,000
6.	Storage space	24
7.	Manpower smoothing	72
	Total	12,732

Table 6–B.   Problem Size for the Linear Programming Formulation

We see that our problem as it is currently formulated has a total of 7,200 variables and 12,732 constraints. We have already noted how the demand constraints could be eliminated from our formulation because they were redundant. In a further effort to reduce the number of constraints let us compare the safety-stock (6.14) and the shelf-life (6.16) constraints. You should notice three things:

1.   the left-hand sides are identical;

2.   the right-hand side of the shelf-life constraints is never less than the right-hand side of the safety-stock constraints;

3.   the shelf-life constraints are of the $\leq$ form and the safety-stock constraints are of the $\geq$ form.

The shelf-life constraints are placing an upper limit on production for the first m months while the safety-stock constraints establish a lower limit. As a result we have

$$\begin{array}{ccc} \text{Safety-Stock} & & \text{Shelf-Life} \\ \text{Constraint} & \leq \sum_{j=1}^{m} \sum_{r=1}^{R_j} x_{irj} \leq & \text{Constraint} \\ \text{Right-Hand Side} & & \text{Right-Hand Side.} \end{array}$$

The expression $\sum_{j=1}^{m} \sum_{r=1}^{R_j} x_{irj}$ can be bound by the upper

limit or the lower limit, but never both limits at the same time.

Some computer programs of the Simplex method take advantage of the fact that only one of these limits may be binding in the optimal solution. Such computer programs allow the user to indicate which constraints are of this form and then the program automatically selects that bound which constraints the solution. The computer program used to solve the candy plant problem did have this capability. Hence, we were able to combine the safety-stock and shelf-life constraints into one set of upper and lower bound constraints thus effectively reducing the number of constraints in our problem to 6,732. This provided a significant savings and is one method that should be considered for effectively reducing the number of constraints in all large-scale linear programs.

6.3   Implementation—Data Preparation

Most inexperienced system analysts or operations researchers assume that once they have defined and formulated the problem the project is almost complete. These people tend to believe that the data preparation process is rather simple and hence can be left to keypunch and computer operators. Actually, especially when we are dealing with a large-scale system, this assumption could not be further from the truth. By studying carefully the next few sections, you will gain a better appreciation and understanding of some of the problems encountered in transforming a linear programming formulation into a meaningful report for the decision-maker. In some respects, the real headaches associated with a large-scale linear programming model just begin with the data preparation phase.

Assuming that we are satisfied with our linear programming formulation, we are now faced with the problem of preparing the input data required in order to set up the initial Simplex tableau. We must prepare the following three types of input data:

1. coefficients of the decision variables for the left-hand sides of the constraints

2. coefficients for the right-hand sides of the constraints

3. objective function coefficients for the decision variables.

We do not have to prepare any information involving slack, surplus, or artificial variables, since the computerized version of the Simplex method will automatically define these variables when setting up the initial Simplex tableau.

Let us start by considering what one encounters when preparing the matrix that contains the coefficients of the decision variables for the left-hand sides of the constraints. For ease of description, we will always refer to this matrix as the data matrix for our linear programming model. Take out a large sheet of paper because we have a data matrix with 6,732 rows and 7,200 columns. Our 6,732 x 7,200 data matrix has a total of 48,470,400 elements. Obviously, this matrix is too large to prepare by hand; we are going to have to design a computer procedure that will generate this matrix for us.

For extremely small problems, such as the common example problems found in linear programming texts, it may be entirely possible for the analyst to write down the entire data matrix and have a keypunch operator prepare the data cards for the linear programming computer program; however, as the size of the linear programming problem increases, the data preparation time and the potential for clerical and keypunch errors makes a computer generated data matrix desirable. Even for a problem as small as 50 variables and 25 constraints, we have a 1,250 element matrix. The time to prepare the data matrix for this moderate-size problem and the possibility of errors in the preparation make the computer generated data matrix a necessity for large-scale models.

Let us now consider what is required in order to prepare the objective function and right-hand side coefficients. If we recall the general expression for our objective function, equation (6.31), and some of the more complex expressions that serve as right-hand sides of our constraints (equations (6.14) and (6.18)) we see that computer generation of these objective function and right-hand side coefficients is the only practical method for developing these data items. Why? Consider the problem of computing 7,200 objective function coefficients of the form of equation (6.31), and 6,732 right-hand side coefficients some of which are the form of equation (6.18). Thus, the approach we will take is to computer generate all necessary input data coefficients.

Data Base

Before we get into the actual process of computer generating the data matrix, right-hand side coefficients, and objective function coefficients, it is important for us to determine the information about the production process that will have to be compiled before these coefficients can be generated. Basically, we must collect information about the known characteristics or input parameters of our problem. This information, which we refer to as the data base, includes items such as the following: production times, production costs, labor rates, safety stock requirements, etc. An outline of the requirements for our data base is shown in Table 6–C.

As you can see, the data base essentially captures the description of the entire production operation. Any detail about the production process that is needed by the linear programming model is available from the data base.

Since the data base is the starting point for developing the input data coefficients, and since we have decided that these coefficients will be computer generated, it is imperative that the data base be in a form that can be directly interpreted by the computer. In the candy plant project, the data base was placed on approximately 3,000 computer cards. An example of the data base card format for the production capacity information is shown in Figure 6–A.

1. General Data
   a. Number of products and identification of each
   b. Number of labor grades and identification of each
   c. Number of pieces of equipment and identification of each
   d. Number of storage types and identification of each

2. Production Capacity Constraint Data
   a. Hours available for each piece of equipment each month
   b. Equipment used on each route for each product
   c. Time required for each unit on each piece of equipment

3. Manpower Capacity Constraint Data
   a. Man-hours of each labor grade available each month
   b. Man-hours of each labor grade required to produce one unit of each product on its specified route

4. Safety-Stock and Shelf-Life Constraint Data
   a. Forecasted demand for each product in each month
   b. Beginning inventories for all products
   c. Inventory life for each product

5. Storage Space Constraint Data
   a. Cubic feet available for each kind of storage each month
   b. Cubic feet of storage type k required by each product

6. Manpower Smoothing Constraint Data
   a. Number of man-hours worked per employee per month
   b. Actual number of employees in each labor grade during the month prior to the schedule
   c. Maximum allowable employee increase and decrease for each labor grade each month

7. Objective Function Data
   a. Cost of producing one unit of each product via each route
   b. Inventory carrying cost for each product

Table 6–C.   Data Base Information for the
Production Scheduling Problem

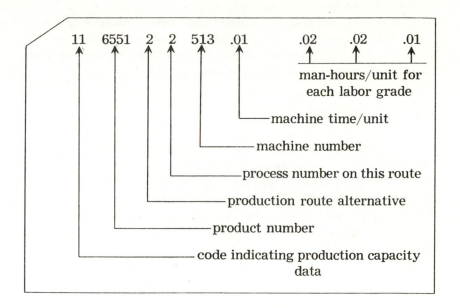

Figure 6–A.   Example Production Capacity Card from the
Data Base

Regardless of how the information in the data base is physically stored, it is extremely important that it be possible to change elements in the data base relatively easily. The reason for this is that an up-to-date data base is the starting point for our solution procedure. In situations where the linear programming model is to be solved on a routine basis, it is important that the efforts required to update the data base be minimal. Actually it is a rare situation when the entire data base has to be revised in order to find a new solution or schedule. In our production scheduling project, the major data base revisions consist of inputting the most recent forecasts for all products (500 cards) and the current actual inventory levels for all products (500 cards). These data are made available from the forecasting department and the warehouse inventory records on a monthly basis.

The remaining two-thirds of the data base is rarely changed from month to month. Most of the information in this remaining two-thirds concerns product routing, machine data, etc. This section of the data base is changed only when products or equipment are added or deleted from the production system. In any case, the use of a punched card data base makes it relatively easy for one individual to replace sections of the data base.

Coefficient Generator

We are now ready to consider the procedure whereby we use our data base to computer generate the data matrix, right-hand side coefficients, and objective function coefficients. To generate these coefficients we must have a specially designed computer program that contains appropriate instructions for accessing the data basé and then performing all necessary calculations required to compute all ceofficients. In essence, the appropriate instructions are computer statements that describe our complete problem formulation. For example, one set of computer instructions describes how to compute the coefficients for the production capacity constraints, another set describes how to compute the coefficients for the manpower capacity constraints, etc. The overall process of calculating these input coefficients is depicted in Figure 6–B.

Figure 6–B.   Coefficient Generation Process

Note that magnetic tape is the output medium. Magnetic tape output is recommended because the number of coefficients to be generated in large-scale systems is extremely large. While a disk storage file is also acceptable, the cost of storing information on disk is necessarily higher. Output on computer cards is usually undesirable when you consider the fact that we may be generating several million coefficients.

What is the coefficient generation program really like? In answering this question, we point out that it is a special purpose program that is tailored to one particular problem. The coefficient generator we have developed here handles only linear programming formulations of the Big Sweet Candy scheduling problem. Other linear programming problems must have their own specially written coefficient generator.

The coefficient generation program may be written in any computer language. Some software manufacturers may have a special coefficient generator language as part of their total linear programming package. In any case, regardless of the language, it is the responsibility of the analyst to work with a computer programmer to design the program that transforms the data base and problem formulation into the required input coefficients. This is not necessarily an easy phase of the implementation process. Actually, it may take more time and create more difficult problems for the analyst than any other phase.

Now let us consider the following question: What does the output from the coefficient generator look like? That is, what is the form of the coefficients as they appear on the output tape? Most often, your linear programming computer code will restrict the output form of the coefficients. Thus, it is important to check the exact form that will be required by your linear programming code before developing the coefficient generator. For example, some codes do not require that zero coefficients be defined. This, of course, can significantly reduce the number of records that must be created by the coefficient generator program.

Our coefficient generator prepares the coefficient output tape such that a list of all elements in the data matrix appears first, followed by a list of right-hand side coefficients, and then a list of objective function coefficients. Only nonzero coefficients are listed on the output tape. As required by our linear programming computer code, each element in the data matrix is written on the output tape in the following form: (1) row position, (2) column position, then (3) coefficient value. Each right-hand side coefficient is written by first specifying the row number followed by the coefficient value; each objective function coefficient by specifying the column number followed by the co-efficient value.

As a result of our linear programming code requirements, it was necessary for us to devise a scheme to name the rows and columns of the data matrix. While we could have used a sequential numbering system such as constraints 1, 2, ... 6732 and variables 1, 2, ... 7200, it would have been extremely difficult to identify exactly what each constraint and what each variable referred to. For example, if at a later date we wanted to modify an element in the constraint for labor grade 2 in the 5th month, we would not have known which constraint (#139?, #6255?, etc.) we needed to alter unless we had taken the time to prepare a master list that could be used to identify each constraint and variable. Actually, even if we had a master list, we would not want to sort through the entire list each time we needed to identify a constraint or variable. Therefore, we decided to name the constraints and variables in such a manner that the name would automatically identify the constraint or variable.

Suppose that we gave the following name to our labor grade 2 in month 5 constraint:

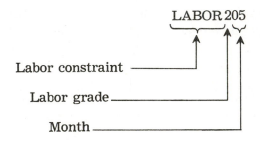

With this scheme, any changes or references to this constraint could be made by simply checking the matrix row LABOR205. Using this notation, what would LABOR311 refer to? Your answer should be the constraint on labor grade 3 in month 11. In a similar manner, we can name the variables, or columns. For example, consider the following production variable name:

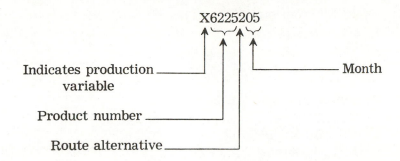

Using this descriptive naming approach, we could identify the coefficient in the row named LABOR205 and the column named X6225205 by writing on our coefficient output tape the row and column name followed by the computer generated coefficient value. For example, if the computer generated coefficient value corresponding to the above row and column was .09, the coefficient generator program would write on the output magnetic tape the following record:

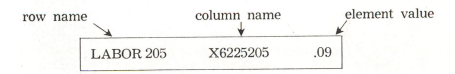

The above record, including the name assignments, is automatically prepared by the coefficient generator program using a list of row and column names as defined by the analyst. Table 6–D contains a list of the row and column names used in the candy plant production scheduling problem.

The above constraint and variable naming capabilities, which are highly desirable, are not available with all linear programming computer codes. If this feature is not available, the constraints and variables will have to be numbered sequentially and a master list of row and column definitions prepared. This is not a serious limitation for small problems, but it is very inefficient with large-scale problems.

A. Production Capacity—EQaaaabb
    1. EQ refers to equipment constraints
    2. Field a contains the machine number
    3. Field b contains the month

B. Manpower Capacity—LABORaab
    1. LABOR refers to manpower constraints
    2. Field a contains the labor grade number
    3. Field b contains the month

C. Safety-Stock and Shelf-Life—INaaaabb
    1. IN refers to inventory constraints
    2. Field a contains the product number
    3. Field b contains the month

D. Storage Space—STOREabb
    1. STORE refers to storage space constraints
    2. Field a contains the storage type number
    3. Field b contains the month

E. Manpower Smoothing—MANSMabb
    1. MANSM refers to manpower smoothing constraints
    2. Field a contains the labor grade number
    3. Field b contains the month

F. Production Variables—Xaaaabcc
    1. X refers to a production variable
    2. Field a contains the product number
    3. Field b contains the product route
    4. Field c contains the month

Table 6–D.   Constraint and Variable Names
             for the Initial LP Matrix of
             the Candy Plant Production
             Scheduling Problem

6.4   Implementation—Computer Solution

Using our computer generated coefficient values, we are now ready to proceed with the Simplex method in order to find an optimal solution to our linear programming model. The solution procedure for virtually all linear programming problems, regardless of size involves a computerized version of the Simplex method. That is, in today's computer world, it is very unusual for even small linear programming problems to be solved by hand calculation. With large-scale problems, the number of hand calculations and the many possibilities for error make a computer solution routine necessary.

For reasonably small problems (less than 250 variables and 80 constraints) the computer program presented in the Appendix of this text will work quite nicely. However, for large-scale linear programming problems, the analyst will find it necessary to use one of the large high-speed computer codes specifically designed to handle large-scale problems. One good large-scale linear programming computer package is IBM's mathematical programming system (MPS or MPSX). These systems have highly efficient linear programming and data handling capabilities that make the solution of a large-scale problem relatively easy. The analyst will, however, need to spend substantial implementation effort in understanding the operation of the linear programming system and determining how his problem should be set up to meet the format and data handling requirements of the system; e.g., the coefficient input format required by the linear programming system. The computer solution process is shown schematically in Figure 6–C.

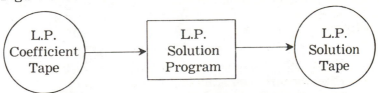

Figure 6–C.   The Computer Solution Step of the
Implementation Procedure

Computer Solution Times

Realizing that we are dealing with a problem having several thousand variables and several thousand constraints, we can see that there will be many calculations required to perform one iteration of the Simplex method and that in searching for an optimal solution we may have to make several thousand iterations. Fortunately the computer can perform these calculations for us; however, even with its efficient data handling and computational procedures, substantial amounts of computer time will be required. Using the MPS package and the IBM 360/65 system, solution of the candy plant scheduling problem required 15 to 20 hours of computer time. Anywhere from five to seven hours were spent on direct iteration computations while the rest of the time was usually spent on data handling or file input/output. Similar size problems solved with the MPSX package (extended version of MPS) indicate these times can be reduced to approximately one-third of the above values. In any case, regardless of the system, the solution of large-scale linear programs requires substantial resources in terms of computer size and computer time.

Reducing Computer Solution Times

With large-scale problems, which almost always require substantial amounts of computer time, the analyst must look for ways to reduce or eliminate some of the computer implementation problems.

One of the first computer solution problems that the analyst will be faced with is that caused by computer down-time. That is, since the large-scale linear programming problem can take several hours of computer time to reach a solution, any computer breakdown during the computer solution stage may cause a loss of all computations that have been completed at that point in time. If this happens, one alternative for the analyst is to return to the data base and data generator, create anew all input coefficients, and then start the Simplex method again. This second computer run will also require several hours of computational time and like the first computer run will be subject to computer failure.

With large, complex high-speed computer systems, it is not unlikely that the computer will experience down-time problems rather frequently. During these down times, although they are sometimes very short in duration, there is always the possibility of destroying all linear programming computations performed prior to that point in time. Besides the frustration of not being able to obtain the complete computer solution on time, the analyst must be concerned with the additional computer time and cost that will be required to resubmit the same program several times.

The above problem can for all practical purposes be eliminated if the analyst periodically saves one of the in-process solutions. That is, the analyst should incorporate a computer procedure that will interrupt the linear programming algorithm periodically—every 100 hundred iterations for example—and record the current solution on magnetic tape. These in-process solutions are of course nonoptimal and are really of no immediate value to the production scheduler; however, if the computer does go down, the most recent in-process solution which has been saved will be much nearer to the optimal solution. Thus, when the computer starts up after a breakdown, the analyst can then insert the most recently saved in-process solution and continue calculations from this point. This type of in-process solution saving procedure provides a safeguard against computer failure and thus makes the computer implementation process more efficient. This in-process solution saving procedure is shown schematically in Figure 6–D.

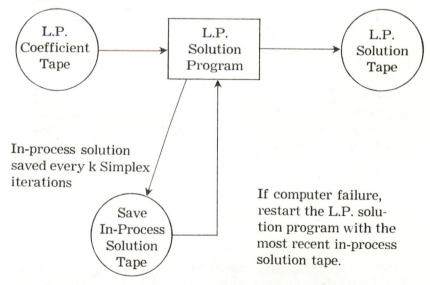

Figure 6–D.   The Computer Solution Process with the Feature for Saving In-Process Solutions

The computer implementation process that we have described so far always starts with the data base being transformed into the initial Simplex tableau. We can think of the initial basis for the Simplex tableau as consisting of the slack and/or artificial variables. While there is nothing wrong with this starting point, it is frequently a starting solution that is far from the optimal solution. Thus, we will experience many iterations and substantial amounts of computer time before we reach the optimal solution. An obvious approach to reducing the computer solution time is to look for ways to obtain a starting solution that is closer to the optimal solution than the initial slack and/or artificial variable solution. The question of course is how does the analyst find a starting point closer to the optimal solution?

If a large-scale linear programming problem is to be solved on a routine time schedule (e.g., every month), the determination of good starting solutions may not be that difficult. For example, suppose that we generate an optimal production schedule in the month of October. You will recall that even though we establish the solution in October, the optimal production schedule shows how many units of each product must be produced in each month throughout the next twelve month period. Suppose that our optimal solution generated in October shows that 500 units of product number 3067 should be produced in the month of May. Suppose further that we are now ready to generate a twelve-month production schedule starting in November. What starting solution would you suggest for product 3067 in the month of May? While admittedly there will probably be revisions in the data base in November (i.e., revised sales forecasts, new products added, etc.), the modifications in the linear programming problem should be relatively minor. Thus, a May production of 500 units of product number 3067, while not necessarily the optimal solution, is probably much closer to the optimal solution than a May production of zero units which is what is implied by an initial tableau consisting entirely of slack and/or artificial variables.

Thus, we are suggesting that if the linear programming problem is to be solved as a routine procedure a starting solution based on the previous period's optimal solution should be closer to the new optimal solution than an initial tableau consisting of slack and/or artificial variables. Hence, computer solution time can be reduced if the basis of the optimal solution for the previous solution is used as the starting solution for the current problem.

It is beyond the scope of this text to go into the procedure of how the previous month's optimal solution basis is merged with the initial tableau of the current month's problem. What we are suggesting here is that this procedure should be considered as a way of reducing computer solution time. Our experience with the candy plant production scheduling problem showed that starting the linear programming algorithm with the previous month's optimal solution resulted in a total computer solution time of approximately ⅓ to ¼ of the time required to solve the same problem starting from an initial tableau with a basis consisting of slack and artificial variables. Hence, the savings achieved by using the previous month's basis was significant.

A final alternative for reducing computer solution time is to consider ways in which the large-scale problem may be broken down into smaller problems. While these smaller problems are still large, each one separately will have less data handling and less computer time requirements. This process of splitting one large-scale linear program into several smaller problems cannot of course be done on an arbitrary basis; the subproblems must be completely *separate* problems. This means that all variables and constraints may appear in one and only one subproblem. For the candy plant project, it was possible to break the large-scale problem down into two subproblems defined by the two plants: Howell Avenue and Fairfield Road. Each plant produced a separate set of products and had separate manpower, machine, and storage capacities. In effect, the two separate plants required the solutions to two completely independent linear programming problems. Instead of the one large 7,200 variable and 6,732 constraint problem, the solution procedure involved solving two separate problems the largest of which contained approximately 3800 variables and 3600 constraints. The computer solution procedure was the same for both subproblems except for the fact that the input data base was specifically prepared for the subproblem being solved.

What advantages does this procedure provide? First, while the total computer time required for solving the two subproblems may be close to the total time necessary to solve the one larger problem, we have at least reduced the computer time and storage requirements of each individual computer run. Thus, our smaller problems have a faster turnaround time, less computer storage problems, and a high probability of completing a run without a computer down-time interruption. Furthermore, if conditions change at one plant—revised forecasts for example—only the particular subproblem affected will have to be solved again. A further advantage of this subproblem approach is that eventually we may find large-scale problems that are just too large to solve with existing software and computer capabilities. Smaller sub-problems may, however, fall within these solution capabilities.

6.5   Implementation—Report Generation

If all has gone well, at this point in the implementation process our computer algorithm will have generated a magnetic tape containing the optimal solution to our linear programming problem. However, even though the optimal solution has been obtained, we are now faced with a significant new problem—that of transforming the information on the magnetic tape into easily readable management reports. As you will see in the discussion that follows, this is another seemingly "trivial" implementation step that can provide additional headaches for the analyst.

One possible approach to this problem is to just print out the linear programming solution tape. A drawback to this approach is that the solution tape contains the variable and constraint names in the form specified by the data generator. For example, using this simple approach, one element of the optimal solution might appear on the printout as follows:

X5411207   175

Recalling the notation for variable names presented in Table 6–D, we see that this element of the optimal solution indicates that 175 units of product 5411 are to be produced on route 2 during the 7th month of the production schedule. Obviously, a solution printout in this form could not be easily interpreted by the managers and/or production control personnel who need the results. This is especially true when you realize that the printout of the entire solution would involve a list of 7,200 variables having the above format. Thus, the analyst must take additional steps to transform the information on the solution tape into readable output reports.

To do this we will have to identify the additional output reports the managers would like to see. For example, what information would they like about machine utilization, manpower levels, inventory levels, etc. All that we have discussed above is the optimal solution in terms of the production volumes for each product. But contained in this production schedule are the utilization data that might also be helpful to the manager.

For the candy plant problem the production control manager specified a need for the following reports: (1) a production schedule for each product each month, (2) a monthly machine utilization report that showed the hours scheduled and the excess capacity for each piece of equipment, (3) a labor summary report which showed the number of people that must be employed each month, and (4) an inventory report which summarized the space requirements in each storage area each month.

While some of the above report items are available as part of the optimal solution to our linear program, other items in the report will require some additional computations. The production schedule can be generated by merely printing out in a more readable format the production variables. Items such as the excess capacity available on the various machines can be obtained by printing out the values of the slack variables. Information such as machine and labor utilization can be found in two different ways. The first approach involves returning to the production schedule and computing the machine and manpower requirements for all items produced in a given month. A quicker and easier computational procedure is to store the machine and manpower capacity figures—the right-hand sides of our original formulation—and then compute the workload or utilization by subtracting the slack variables—the unused capacities—from the right-hand side values.

Surely you can now see that the report generation step can be a rather complex process. The complexity will depend primarily on the needs of the manager and how the LP solution tape data has to be manipulated in order to obtain this information.

Because of the complexity of report generation, a special purpose report generation program is usually required. This report generation program can be written in a special report language designed for a particular linear programming code, or it may be written in one of the general programming languages. The report generation program can be stored separately on disk or magnetic tape until the optimal LP solution has been reached. The LP solution tape can then interface with the report generation program to provide the final managerial report. This process is shown schematically in Figure 6–E.

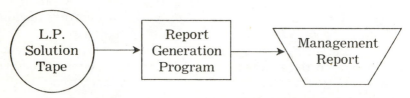

Figure 6–E.   The Report Generation Process

A portion of the production schedule report generated from the candy plant scheduling problem is shown in Table 6–E. A portion of the machine utilization report for the same study is shown in Table 6–F.

Production Schedule Report
(cases)

Product Number	Route	Jan.	Feb.	Mar.	Apr.
1250	1	—	200	300	—
1250	2	—	200	—	—
1270	1	100	150	100	100
1285	1	—	—	—	400
1295	1	300	100	—	—
1295	2	—	—	—	—

Table 6–E.   A Portion of the Production Schedule
            Report Generated in the Candy Plant
            Scheduling Problem.

All Figures in Hours

Machine Number		Jan.	Feb.	March	April
5001	Available	40.0	36.0	40.0	38.0
	Scheduled	38.2	36.0	36.0	32.0
	Excess	1.8	0.0	4.0	6.0
5002	Available	40.0	20.0	20.0	20.0
	Scheduled	29.2	20.0	15.5	13.0
	Excess	10.8	0.0	4.5	7.0

Table 6–F. A Portion of the Machine Utilization Report Generated in the Candy Plant Scheduling Problem

6.6 Avoiding Infeasible Solutions

If all has gone well up to this point, we now have a complete linear programming solution and a managerial report containing information about the optimal production schedule. However, as we attempt to solve problems with a large number of variables and a large number of constraints we must be prepared for cases where a subset of the constraints cannot be simultaneously satisfied. That is, we may encounter situations where there is no feasible solution to our linear programming problem. With the large number of constraints found in large-scale problems it is nearly impossible for the analyst to know in advance if a feasible solution does in fact exist.

As an example of how infeasibility might occur, consider the case where we have a product with an extremely high demand and at the same time an extremely short shelf-life. Since the shelf-life is short, we can not produce the product very far in advance of when it will be needed. Now let us suppose the machine capacity available for this product in the month it is needed is small. If such a case as this existed, it would be impossible to satisfy both the shelf-life restriction and the machine capacity restriction simultaneously. To the production scheduler this simply means that, given the manpower and machine availabilities, he has placed too many requirements (such as short shelf-life) on the system to make an acceptable production schedule possible.

Of course, the really serious disadvantage of infeasible solutions is that our linear programming solution procedure will not generate the schedule. That is, our linear programming algorithm will simply indicate that there is no feasible solution and stop. The output from the linear programming solution program will usually be a statement indicating a feasible solution does not exist, and possibly an indication of the constraints or conditions that are causing the infeasibility. This output from the system will be completely unacceptable to the production scheduling people especially when they are impatiently awaiting a computer run that will provide the information for the next week's or next month's schedule.

If we are in fact faced with the case of no feasible solution, we will have to track down the cause of this infeasibility and then investigate ways in which the initial problem can be modified, if possible, to guarantee a feasible solution. Of course, even if we are able to modify our problem such that a feasible solution is obtained there is nothing that will guarantee us that the next time the linear programming model is solved we will obtain a feasible solution. It is entirely possible that by changing a few input conditions infeasibility may result from a completely different set of constraints for a completely different set of reasons.

We can easily see that many problems will arise as management and production control personnel become impatient and disenchanted with a production scheduling procedure that does not always provide a feasible solution. Therefore, the wise analyst will look for ways in which he can guarantee that the problem will always provide a complete managerial report with a solution as close to the optimal solution as possible. If conflicting constraints do exist, and hence there is no feasible solution, the system should be designed to prepare a summary report for the manager indicating where and how much extra manpower and/or machine time must be found in order to obtain a feasible solution. With this type of information, the manager can then make alternative decisions such as using overtime to meet the schedule requirements.

Now that we have identified some of the problems associated with infeasible solutions and pointed out the need for avoiding infeasibility, we are ready to identify ways in which a feasible solution and near optimal production schedule can be guaranteed. To accomplish this we will need to modify our original problem formulation somewhat. First, let us consider the production capacity constraints as specified by equation (6.2).

$$\sum_{i=1}^{N} \sum_{r=1}^{R_i} t_{ire} x_{irm} \leq C_{em} \qquad (6.2)$$

where

t_{ire} = production time in hours required to produce product i via route r on equipment e

x_{irm} = units of product i produced via route r in month m

C_{em} = production hours available for equipment e in month m.

Suppose that our total production time as specified by the left-hand side of equation (6.2) is forced by other constraints to be greater than the available number of production hours, C_{em}. If this was the case, the above constraint would in effect cause an infeasible solution and the computer algorithm would stop after identifying the specific equipment e and month m that were causing the infeasibility condition. A modification in the problem formulation that will allow the Simplex method to continue until an optimal solution to the modified problem is reached is as follows:

Let C'_{em} = an excess capacity variable indicating the extra capacity required by equipment e in month m in order to obtain a feasible solution.

Thus, our production capacity constraint becomes

$$\sum_{i=1}^{N} \sum_{r=1}^{R_i} t_{ire} x_{irm} \leq C_{em} + C'_{em}$$

Placing the variables on the left-hand side, we get

$$\sum_{i=1}^{N} \sum_{r=1}^{R_i} t_{ire} x_{irm} - C'_{em} \leq C_{em}. \qquad (6.32)$$

Of course, the extra capacity C'_{em} is an additional variable that we will want to enter the solution only in order to obtain a feasible solution. To make sure this variable is used only when necessary, we assign an extremely high cost to the variable when it appears in the objective function (e. g., the magnitude of this cost co-efficient could be of the order of 1000 times the highest other cost coefficient in the objective function). By using a similar extra capacity variable on the manpower capacity constraint (equation (6.3)) and the storage-space constraints (equation (6.18)), we can guarantee that the linear programming algorithm will run to completion and that the production schedule report will show production quantities in each month that will satisfy the demand, safety-stock, and shelf-life conditions. Note that the method of using excess capacity variables to guarantee feasibility need not be applied to all constraints. In the modified formulation suggested above we are allowing some flexibility in the specification of production capacity, manpower capacity and storage capacity in order that the demand, safety-stock and shelf-life constraints may be met exactly.

Now let us see what happens if one of these excess capacity variables appears in the optimal solution. For example, suppose that in order to guarantee a feasible solution we had to have five hours of extra capacity for equipment 5001 in month 7. The manager and the production control personnel now have some very valuable information that can be used to solve this problem in a variety of ways. Specifically, they know that in the seventh month they will need five extra hours of capacity on machine 5001 if they are going to meet the demand, safety-stock, and shelf-life conditions that they have specified. An initial alternative might be to schedule that machine for overtime production during that month. On the other hand, by looking at the machine utilization report the manager might note that this equipment was not used to capacity in the previous month. This might lead him to an investigation of the shelf-life conditions for some of the products produced on this equipment to see if earlier production might be acceptable. In any case, the management report should include a section that identifies the amount of the extra production capacity, extra manpower capacity, and/or extra storage capacity needed. The value of preparing this section of the report is that the manager will have this information prior to the month in which the trouble occurs. He can then use his own judgment as to the best way of avoiding the difficulty.

6.7   A Review of the Development and Implementation Procedure

Let us reflect for a moment on what we have discussed thus far. We have defined the five steps involved in the development and implementation process for a linear programming application: problem definition, problem formulation, data preparation, computer solution, and report generation. Problem definition, we saw, required a good deal of ingenuity and thought before we could decide exactly what the objectives and constraints of our problem really were. The problem formulation stage requires someone who is skilled in the art of transforming the well-defined problem into the mathematical form required for a linear programming model. We have shown that once the problem has been formulated, the actual useable solution may still be quite a ways off. The next step involves preparing data for the entire problem. If we are talking about a large-scale problem it will be necessary to prepare and update on a continual basis an extremely large data base. This data base must be transformed by a specially designed computer program into the coefficients required to set up the initial linear programming tableau. Substantial time and computer skills must be devoted to this step.

The fourth step is the preparation of a computer solution. The performance of this step requires the assistance of someone skilled in the operation of computer systems and someone specifically trained to handle large-scale linear programming software packages. We may encounter some problems at this stage due to software or hardware limitations of the computer system. The final implementation step is the preparation of a managerial report containing a variety of pertinent information about the production schedule. We have seen that this report must be specifically designed for the problem being solved and the specific needs of the manager. This step will also require computer programming skills in order to prepare the report from data that is generated by the computer solution. In total, the development and implementation processes represent a sizeable task. To be successful, we must have the ingenuity for the problem definition stage, the mathematical training at the problem formulation stage, computer programming skills at the data preparation stage, computer systems skills at the computer solution stage, and computer programming skills at the report generation stage.

6.8   Application of the Results

The final value of a large-scale linear programming model, such as the one defined in this chapter, must be measured in terms of the value it provides in terms of solving a real problem. This value may be based on the fact that without the model the problem is just too complex to solve, or it might be based on the fact that the computer solution procedure provides a quicker and easier way to solve the problem.

Let us look specifically at the applicability of our production scheduling solution. First, the major contribution is that we have a production schedule for each of the next twelve months. The production scheduling personnel can now use this information to generate specific week by week schedules for the short-run period. In addition, we point out that by reviewing the scheduled shifts on each machine, the scheduled labor, and the scheduled storage, the production schedulers will be able to identify when machines, types of manpower, and/or storage areas may become overloaded. By having this information usually far in advance, the scheduler may be able to take action that will avoid potential overload conditions in the future.

A further major advantage of this model is that it can show the manager how changes in product mix, equipment, labor, safety-stock, inventory cost, etc., will affect his operation. These changes can be evaluated before they ever actually take place. For example, suppose the manager is considering installing a new special piece of equipment. Before the equipment is ever purchased we can modify the data base to include that piece of equipment. By solving the resulting linear programming problem and analyzing the projected schedule, the manager will be able to observe the effect of the new equipment before it is purchased. Hence, the projected utilization and cost estimates can be important inputs into future equipment and/or manpower level decisions.

6.9 Summary

We have presented in this chapter many of the aspects of actually implementing a large-scale linear programming model. While the specific example discussed involved production scheduling, the general implementation problems and principles are the same that might be encountered in a variety of large-scale problems. In this regard, we saw that model development and implementation was a five-step procedure: problem definition, problem formulation, data preparation, computer solution and report generation.

One factor that we should not overlook is the time and cost associated with the project. First of all, the development cost is not cheap. We are talking about using the services of rather highly trained and highly paid personnel. In addition to the cost of the analyst's time, the firm will also experience computer development time costs. In addition, you should be aware of the fact that most novices in this area will tend to greatly underestimate the time required to get the system operational. The computer testing and debugging phases will take substantial time. In the final analysis, the manager must decide whether or not the advantages of such a system outweigh the development time and cost requirements. The convenience and savings obtained by a well designed system can often more than pay for the large implementation time and cost requirements.

6.10 Glossary

1. *Large-scale linear programming model*—A model with a large number of variables and/or constraints.
2. *Model development*—Defining the real-world decision making problem and formulating an appropriate linear programming model.
3. *Model implementation*—Includes data preparation, computer solution, and report generation.
4. *Production capacity constraints*—Constraints that limit the amount of production because of available machine time.
5. *Manpower capacity constraints*—Constraints that limit the amount of production because of available man-hours of labor.
6. *Demand constraints*—Constraints that guarantee demand is satisfied.

7. *Safety-stock constraints*—Constraints that force production to be greater than forecasted demand in order to meet unexpected demands.

8. *Shelf-life constraints*—Constraints that insure products will not remain in inventory longer than their maximum inventory life.

9. *Storage-space constraints*—Constraints that insure that ending inventories do not exceed the available storage space.

10. *Smoothing constraints*—Constraints that attempt to maintain the same level of resource activity in two adjacent periods (i. e., manpower level, production level, etc.).

11. *Data matrix*—A matrix of computer generated decision variable coefficients for the left-hand sides of the constraints.

12. *Data base*—All data or information required to define the linear programming problem.

13. *Coefficient generator program*—A specially designed computer program that extracts information from the data base in order to create the (1) data matrix, (2) right-hand side constraint coefficients, and (3) decision variable objective function coefficients.

14. *Report generator*—The specially written computer program that transforms the linear programming solution data into readable management information reports.

6.11 Problems

1. What are the five major steps associated with developing and implementing any linear programming model? Which steps are often erroneously considered trivial?

2. Discuss how in most production scheduling problems demand and safety stock constraints are redundant.

3. Suppose that in the Par, Inc. problem in Chapter 1, management would like to consider a two-period operation. Also suppose maximum sales are set at 400 Standard bags and 200 Deluxe bags during the first period and 600 Standard bags and 300 Deluxe bags during the second period. If inventory carrying costs are $.50 for each Standard bag and $.75 for each Deluxe bag, formulate the problem as a two-period model that allows inventory to be carried between periods.

4. Suppose in problem 3, it is desirable to use approximately the same work activity level in each of the four departments. Show how department smoothing constraints can be added to this two period problem.

5. Delevop the argument of how the inventory cost (equation 6.25) can be written in terms of production variables x_{irm} instead of ending inventory levels s_{im}. Does the inventory cost expressed in terms of production variables indicate the total inventory cost? Explain.

6. How many data entries are required in the initial tableau of a 50 variable, 40 constraint linear programming problem? Discuss the data handling problems associated with this model.

7. Describe in detail the data base for the Par, Inc. problem. Establish a data card format that might be used to maintain the data base file.

8. Establish a complete variable and constraint naming system for the Par, Inc. problem.

9. Discuss the computer solution problems associated with implementing a large-scale linear programming model. Comment on data handling procedures, computer time problems, etc. What are the two suggested ways for reducing computer time problems?

10. Discuss the steps involved in transforming the linear programming output into meaningful reports. What type reports might be desirable for the Par, Inc. problem? Illustrate the format of these reports.

11. In the Par, Inc. problem, suppose that management required a certain minimum number of each type of bag: $x_1 \geq a_1$ and $x_2 \geq a_2$. If a_1 and a_2 were large enough, it might turn out that not enough resources were available to meet the minimum requirement; thus, infeasibility would result. Show how you would reformulate the problem to guarantee a feasible solution. Define and interpret the meaning of any new variables required. What would the objective function coefficients be for these new variables?

12. What applications might management make of the Par, Inc. linear programming model other than just determining an optimal product mix? Explain.

*

Appendix

LINEAR PROGRAMMING COMPUTER PROGRAM

The purpose of this appendix is to show the reader how to use the computer program in section A.4 to solve linear programming problems. For those readers not familiar with a computer, a computer program is a set of instructions that tells a computer to perform a specific sequence of operations. The computer program included in this appendix is nothing more than a set of instructions that tells the computer how to solve linear programming problems using the Simplex method. The program was designed for ease of student use. To use the program, the student only needs to learn how to prepare a few punched cards that will describe the problem to be solved. Thus, although no computer programming knowledge is required to use the program we shall be describing, it is assumed that the reader is familiar with punched cards and the process of key-punching information onto punched cards.

Since the computer program in this appendix is used to solve linear programming problems, we have named it LP. LP is a computer program written in a programming language called FORTRAN IV. (FORTRAN IV is just one of the many languages that enable programmers to write instructions to a computer.) Thus, LP is a particular set of FORTRAN IV instructions that will enable a computer to solve linear programming problems. A complete listing of LP is given in section A.4.

LP operates upon a set of data in order to produce output results. In our case, the data supplied to LP provides a description of the linear programming problem we want to solve, and the output is a description of the linear programming solution to the problem. Thus, one of the main things we will have to learn how to do is prepare problem description cards. Before examining the specific details of preparing the input data, however, let us get a good look at the overall process of solving linear programming problems using LP.

We assume that your instructor will supply you with a punched deck version of LP for you to use every time you want to solve a linear programming problem. In addition, your instructor will

also supply you with some computer system control cards. Because these cards vary among computer installations (even among
those with the same make and model computer), you will have
to obtain instructions as to how the LP program and system
control cards go together. The function of these system control
cards is not important to us here. They serve many purposes
such as identifying the user to the computer in order to check if
the user is authorized to run programs, separating the computer
program from data to the program, etc. The composition of the
deck of punched cards that you must submit to your computer
center will typically appear as shown in Figure A–A. Note that
the problem description cards must be followed by two blank

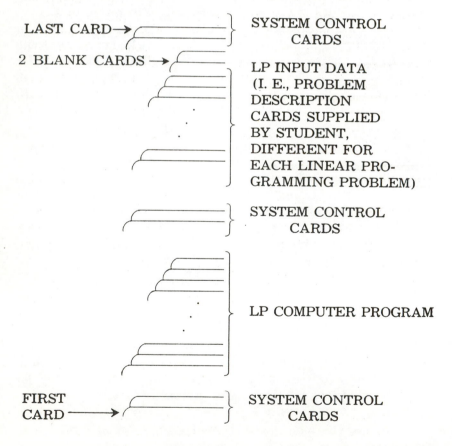

LAST CARD→ } SYSTEM CONTROL
CARDS

2 BLANK CARDS → LP INPUT DATA
(I. E., PROBLEM
DESCRIPTION
CARDS SUPPLIED
BY STUDENT,
DIFFERENT FOR
EACH LINEAR PRO-
GRAMMING PROBLEM)

} SYSTEM CONTROL
CARDS

} LP COMPUTER PROGRAM

FIRST
CARD ——→ } SYSTEM CONTROL
CARDS

Figure A–A.   Typical LP Input Deck

cards.   We will refer to this complete set of cards as the LP Input
Deck.

Let us consider the process that you must follow in order to
solve problems using LP.   First, you begin by formulating the
problem(s) you want to solve.   Your formulation can include
less-than-or-equal-to constraints, greater-than-or-equal-to constraints, and/or equality constraints.   You will not have to worry
about formulating the problem in Standard form, because LP
will do this automatically for you.   Next, using the instructions
below for preparing problem description cards, you must keypunch a set of problem description cards for each problem you
want to solve.   Finally, you put the LP Input Deck together as
shown above and submit the completed LP Input Deck to your
computer installation.   The results from the computer will come
back to you in the form of printed output.   (An example problem
complete with computer output is presented in section A.3.)   This
overall process of solving linear programming problems using the
LP computer code is shown in Figure A–B.

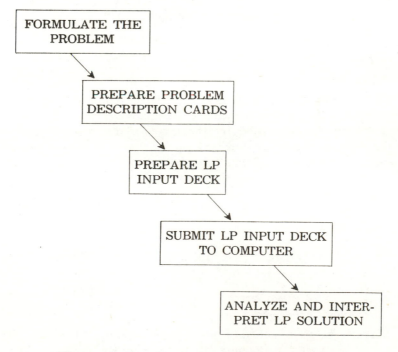

Figure A–B.   Computer Solution Procedure

A.1   Problem Description Cards

For every linear programming problem you want to solve using LP, you must prepare a complete set of problem description cards. There are four different types of problem description cards required for each problem you want to solve. The four types of cards and their functions are listed in Table A–A below.

ORDER	CARD TYPE	FUNCTION
1	User Label Card	Identify Problem
2	Problem Attributes Card	Input General Description of Problem
3	Constraint Input Cards	Input Constraint Coefficients
4	Objective Function Card	Input Objective Function Coefficients

Table A–A.   Problem Description Cards

We will now consider each type of problem description card in detail.

(1) User Label Card

The user label card is the first problem description card for each problem you want to solve. This card simply provides a heading for the computer output. In this way, the output results can readily be identified by the user. This heading may be punched anywhere in columns 1 through 80. If no user label is desired, a blank card *must* be inserted.

LABEL CARD KEYPUNCH INSTRUCTIONS:
ANY MESSAGE IN COLUMNS 1–80

(2) Problem Attributes Card

This is the second card of every problem description set. Five attributes of each problem must be known before any other in-

formation can be input. The five attributes and their functions
are listed in Table A–B.

ATTRIBUTE	CARD COLUMN LOCATION	EXPLANATION
Problem Type	1–4	Keypunch 1 if a maximization problem, 2 if a minimization problem
Decision Variables	5–8	Number of decision variables in problem formulation
Less-Than-Or-Equal-To Constraints ($\leq$ Constraints)	9–12	Number of $\leq$ Constraints
Greater-Than-Or-Equal-To Constraints ($\geq$ Constraints)	13–16	Number of $\geq$ Constraints
Equal-To Constraints (= Constraints)	17–20	Number of = Constraints

Table A–B.   Problem Attributes

When keypunching the problem attributes, you can place the
value of the attribute anywhere within the specified field length;
e.g., if you are solving a maximization linear program, you could
keypunch the problem type attribute of 1 in either column 1,
column 2, column 3, or column 4. If the value of any attribute
is zero, its field may be left blank.

ATTRIBUTES CARD KEYPUNCH INSTRUCTIONS:

EACH ATTRIBUTE IS KEYPUNCHED IN A SEP-
ARATE 4 COLUMN FIELD ACCORDING TO
THE ORDER SPECIFIED IN TABLE A–B.

---

(3) Constraint Input Cards

Constraints must be input in the order shown in Table A–C.

ORDER	CONSTRAINT TYPE		FUNCTION
1	$\leq$ Constraints	Input	$\leq$ Constraint Coefficients
2	$\geq$ Constraints	Input	$\geq$ Constraint Coefficients
3	$=$ Constraints	Input	$=$ Constraint Coefficients

Table A–C.   Constraint Input Order

Only the coefficients associated with each constraint are entered. For each constraint, the coefficient of the first decision variable must be keypunched within columns 1–8, the coefficient of the second decision variable must be keypunched within columns 9–16, etc.   Thus, each coefficient occupies one 8-column field of the punched card.   Hence, the maximum number of coefficients that can be placed on one 80-column card is 10.   Because of this, one constraint may continue onto many cards.   After all the coefficients for the decision variables are entered for a particular constraint, the next 8-column field of the punched card must be

---

CONSTRAINT CARDS KEYPUNCH INSTRUCTIONS:

EACH CONSTRAINT BEGINS ON A NEW CARD IN THE FOLLOWING ORDER: ALL $\leq$ CONSTRAINTS, THEN ALL $\geq$ CONSTRAINTS, THEN ALL = CONSTRAINTS.

EACH COEFFICIENT IS KEYPUNCHED WITHIN ONE 8-COLUMN FIELD.   AFTER ALL DECISION VARIABLE COEFFICIENTS ARE KEYPUNCHED IN THIS WAY, THE NEXT 8-COLUMN FIELD WILL BE KEYPUNCHED WITH THE RIGHT-HAND SIDE VALUE.

THE NUMBER OF 8-COLUMN COEFFICIENT FIELDS USED MUST EQUAL THE NUMBER OF DECISION VARIABLES PLUS ONE 8-COLUMN FIELD FOR THE RIGHT-HAND SIDE VALUE.   ZERO VALUES MAY BE LEFT BLANK.

keypunched with the corresponding right-hand side value. This procedure continues for each constraint in the order specified in Table A–C.

Each constraint must begin on a new card. Read this last sentence again. When keypunching this constraint information, any zero coefficients or zero right-hand side values may be left blank, but the number of 8-column fields used by each constraint must equal the number of decision variables plus one additional 8-column field for the right-hand side value.

(4) Objective Function Card

Preparation of the objective function card is quite similar to preparation of the constraint cards. Each objective function coefficient must be keypunched within an 8-column field. Variables must be in the same order as they were for constraint coefficients. Zero coefficient values may be left blank, but the number of 8-column fields used must equal the number of decision variables.

OBJECTIVE FUNCTION CARD KEYPUNCH
INSTRUCTIONS:

EACH DECISION VARIABLE COEFFICIENT IS KEY-
PUNCHED WITHIN ONE 8-COLUMN FIELD.

A.2 Multiple Program Description Sets

Using LP, the user can solve more than one linear programming problem with each computer run. All that is required is that the user prepare a complete problem description set for each problem to be solved. The label card for each succeeding problem to be solved is placed after the objective function card for the preceding problem description set. A typical setup for solving two problems would appear as shown in Figure A–C. Note that the last set of problem description cards must be followed by two blank cards.

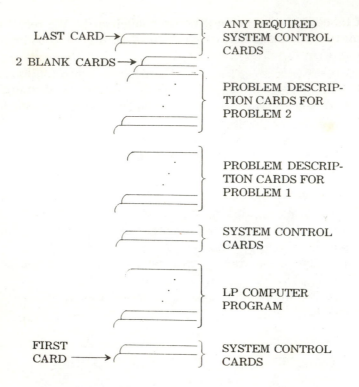

Figure A–C. Typical Multiple Description Setup

A.3   Example—The Par, Inc. Problem

In this section we will describe how to prepare the problem description set for the Par, Inc. problem introduced in Chapter 1 and illustrate the output generated by the LP computer program solution to the problem. In Chapter 1 we saw that the linear programming formulation of the Par, Inc. problem was

$$\max 10x_1 + 9x_2$$

s.t

$$\tfrac{7}{10}x_1 + 1x_2 \leq 630$$
$$\tfrac{1}{2}x_1 + \tfrac{5}{6}x_2 \leq 600$$
$$1x_1 + \tfrac{2}{3}x_2 \leq 708$$
$$\tfrac{1}{10}x_1 + \tfrac{1}{4}x_2 \leq 135$$
$$x_1, x_2 \geq 0.$$

Let us begin by preparing the user label card. We recall that the information placed on the user label card is used to identify the output information generated by LP. For illustrative purposes, assume we wanted to identify our output information with the heading "PAR, INC. EXAMPLE PROBLEM". Following our rules for keypunching a user label card, we prepare the user label card as follows:

COLUMN

```
1                                                           8
                                                            0
┌─────────────────────────────────────────────────────────┐
│  PAR, INC. EXAMPLE PROBLEM                    WWW         │
│                                                          │
│                                                          │
└─────────────────────────────────────────────────────────┘
```

Next, we must prepare the appropriate problem attributes card. This leads us to define values for the attributes as shown in Table A–D.

ATTRIBUTE	VALUE
Problem Type	1
Decision Variables	2
≤ Constraints	4
≥ Constraints	0
= Constraints	0

Table A–D. Attribute Values for Par, Inc. Problem

Following our rules for keypunching a problem attributes card, we keypunch this information in the following manner.

COLUMN

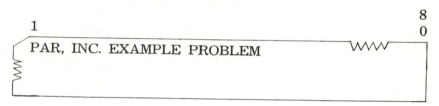

Note that we can place each attribute value anywhere within its four-column field, and that a zero attribute value can be keypunched with a 0 or the field can be left blank.

We now begin keypunching our constraint cards. The corresponding four constraint cards appear as follows:

COLUMN

	1	2
8	6	4
.7	1	630

COLUMN

	1	2
8	6	4
.5	.833	600

COLUMN

	1	2
8	6	4
1	.667	708

COLUMN

	1	2
8	6	4
.1	.25	135

Finally, we prepare the objective function card in a similar manner. Thus, we keypunch

COLUMN

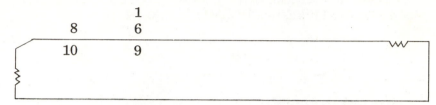

Our problem description set is now complete and ready to be placed together with the LP program. The computer solution corresponding to this LP Input Deck appears as follows:

PAR, INC. EXAMPLE PROBLEM

LP INPUT DATA
****

OBJECTIVE FUNCTION TO BE
MAXIMIZED WITH
2 DECISION VARIABLES

SUBJECT TO:

4 < = CONSTRAINTS (4 SLACK VARIABLES)
0 > = CONSTRAINTS (0 SURPLUS VARIABLES)
0 = CONSTRAINTS (0 ARTIFICIAL VARIABLES)

CONSTRAINT COEFFICIENTS AND RIGHT HAND SIDE
****

CONSTRAINT NO. 1

   0.7000   1.0000

< =   630.0000

CONSTRAINT NO. 2

   0.5000   0.8330

< =   600.0000

CONSTRAINT NO. 3

   1.0000   0.6670

< =   708.0000

CONSTRAINT NO. 4

   0.1000   0.2500

< =   135.0000

OBJECTIVE FUNCTION COEFFICIENTS
****
            10.0000   9.0000

PAR, INC. EXAMPLE PROBLEM

LP SOLUTION
****

OBJECTIVE FUNCTION MAXIMIZED AT   7.66741E 03

3 ITERATIONS REQUIRED

INCLUDED VARIABLES
****

VARIABLE NUMBER	QUANTITY OF THIS VARIABLE	VARIABLE TYPE	ASSOCIATED WITH CONSTRAINT NUMBER
1	5.3984E 02	DECISION	
2	2.5211E 02	DECISION	
4	1.2007E 02	SLACK	2
6	1.7988E 01	SLACK	4

ASSOCIATED DUAL VARIABLES
****

CONSTRAINT NO.	VALUE OF THE DUAL VARIABLE
1	4.37065
2	0.0
3	6.94053
4	0.0

We suggest that you prepare the above cards to familiarize yourself with the above process, and then get a computer solution to the problem.

A.4   Fortran IV Linear Programming Code [1]

LINEAR PROGRAMMING ALGORITHM
   9/15/73

   INPUT PARAMETERS

CARD ORDER	FORMAT	IDENTIFICATION	EXPLANATION
1	20A4	USER LABEL	COLUMNS 1 THRU 80 MAY BE USED FOR USER LABEL. INSERT BLANK CARD IF NO LABEL IS DESIRED.
2	5I4	PROBLEM ATTRIBUTES	THE FIVE ATTRIBUTES NECESSARY FOR INPUT ARE , IN ORDER:

> (1) TYPE 1 FOR MAXIMUM OR 2 FOR MINIMUM
> (2) NUMBER OF REAL VARIABLES
> (3) NUMBER OF <= CONSTRAINTS
> (4) NUMBER OF >= CONSTRAINTS
> (5) NUMBER OF = CONSTRAINTS

3	10F8.0	CONSTRAINT INPUT	(1) CONSTRAINTS MUST BE INPUT IN THE ORDER LISTED ABOVE

> (2) COEFFICIENTS MUST BE TYPED LEFT TO RIGHT WITH CONST-ANT TERM FOLLOWING THE LAST COEFFICIENT
> (3) EACH CONSTRAINT MUST BEGIN ON A NEW CARD
> (4) ZERO COEFFICIENTS MAY BE LEFT BLANK

4	10F8.0	OBJECT FUNCTION INPUT	THE OBJECT FUNCTION MUST BE TYPED LEFT TO RIGHT

NOTE: THIS SAME ORDER MAY BE REPEATED (BEGINNING WITH THE USER LABEL) FOR AS MANY DATA SETS AS REQUIRED.

NOTE: FINAL DATA SET MUST BE FOLLOWED BY TWO BLANK CARDS

THE FOLLOWING DRUM CARD WILL FACILITATE PUNCHING THE DATA.
USE PROG 1 FOR 10F8.0, PROG 2 FOR 5I4.
COL.

```
        1        2        3        4        5        6        7
1       0        0        0        0        0        0        0
5D&DED&D5D&DED&D5D&DEDDD4DDDDDDD4DDDDDDD4DDDDDDD4DDDDDDD4DDDDDDD4DDDDDDD4D
COL.
7   8
5   0
DDDDDD
```

```
          EXAMPLE
************************************

C
C     OBJECT FUNCTION TO BE MAXIMIZED:
C     .04X1 + .05X2 + .08X3 + .06X4 + .03X5 = E
C
C     SUBJECT TO THE FOLLOWING CONSTRAINTS:
C     -X1 -X2 +X3 -X5 <= 0.
C     X1 -X3 -X4 >= 0.
C     X1 +X2 +X3 +X4 +X5 = 100.
C
C
```

--------⦁--------

[1] The authors are indebted to Dean Albert J. Simone, College of Business Administration, University of Cincinnati, for making this code available.

```
C        WILL BE PUNCHED IN THE FOLLOWING MANNER:
C
C                                          COLUMN
C                     1   1   2   2   2   3   3   4   4   4   5   5   6   6   6   7
C          1   4   8   2   6   0   4   8   2   6   0   4   8   2   6   0   4   8   2
C
C CARD
C   1   THIS CARD CONTAINS PROBLEM NAME
C   2       1   5   1   1   1
C   3          -1.         -1.          1.                  -1.          0.
C   4           1.                     -1.         -1.                   0.
C   5           1.          1.          1.          1.          1.     100.
C   6          .04         .05         .08         .06         .03
C
C
C
C
C***********************************************************************************
C
C
C         THE TABLEAU CONSTRUCTED BY THIS PROGRAM IS AS FOLLOWS
C
C                                          COLUMN
C                    ***REAL VAR.** ****FICTICIOUS VAR.****
C          1   2   3   -   -   -   -NR3   -   -   -   -   -   -   N4
C   1   V   O   C   S   S   S   S   S 1. 0. 0. 0. 0. 0. 0. 0.
C   2   A   B   O   S   S   S   S   S 0. 1. 0. 0. 0. 0. 0. 0.
C   3   R   J   N   G   G   G   G   G 0. 0.-1. 0. 1.^0. 0. 0.
CR  -           S   G   G   G   G   G 0. 0. 0.-1. 0. 1. 0. 0.
CO  -   N   C   T   E   E   E   E   E 0. 0. 0. 0. 0. 0. 1. 0.
CW  -   O   O       E   E   E   E   E 0. 0. 0. 0. 0. 0. 0. 1.
C KQ           OB  OB  OB  OB  OB  OB  0. 0. 0.   L   L   L   L
C KS        TC  OC  OC  OC  OC  OC  OC  OC  OC  OC  OC  OC  OC  OC
C N5            NE  NE  NE  NE  NE  NE  NE  NE  NE  NE  NE  NE  NE
C
C***********************************************************************************
C
C        COLUMN 1 CONTAINS VARIABLE NUMBERS
C
C        COLUMN 2 CONTAINS OBJECTIVE COEFFICIENTS
C        COLUMN 3 CONTAINS CONSTRAINT CONSTANTS
C        COLUMNS 4 - NR3 CONTAIN REAL VARIABLE COEFFICIENTS
C        COLUMNS NR3+1 - NR4 CONTAIN FICTICIOUS COEFFICIENTS
C        S DESIGNATES .LE. CONSTRAINT COEFFICIENTS
C        G DESIGNATES .GE. COEFFICIENTS
C        E DESIGNATES EQUALITY CONSTRAINTS
C        L REPRESENTS AN EXTREMELY LARGE NUMBER
C        ROW KQ CONTAINS OBJECT FUNCTION COEFFICIENTS
C        ROW KS CONTAINS TOTAL COST (TC) AND GROSS EVALUATION VALUES
C        ROW N5 CONTAINS NET EVALUATION VALUES
C
C        NOTE:   THE MATRIX CONTAINING THE TABLEAU WILL HAVE THE FOLLOWING
C        DIMENSIONS.
C
C        NO. OF COLUMNS = ( NO. REAL VARIABLES ) + ( NO. .LE. CONSTRAINTS )
C        + ( NO. .EQ. CONSTRAINTS ) + 2*( NO. .GE. CONSTRAINTS ) + 3.
C
C        NO. OF ROWS = ( NO. .LE. CONSTRAINTS ) + ( NO. .EQ. CONSTRAINTS )
C        + ( NO. .GT. CONSTRAINTS ) + 3.
C
C***********************************************************************************
C
C
C        DIMENSION TABLEAU A, VECTOR PROB FOR USER LABEL AND VECTOR Z FOR
C        OUTPUT LABELS
C        THE VECTOR 'B' STORES THE CO-EFFS IN THE OBJECTIVE FUNCTION.
C
         DIMENSION PROB(20), Z(15)
         DIMENSION A(84,252),B(84)
C
         DATA Z(1),Z(2)/'MAX','MIN'/
         DATA Z(3),Z(4)/'SLAC','K   '/
         DATA Z(5),Z(6)/'SURP','LUS '/
         DATA Z(7),Z(8)/'DECI','SION'/
         DATA Z(9), Z(10), Z(11)/ 'ARTI', 'FICI', 'AL  ' /
         DATA Z(12)/ '    '/
         DATA Z(13),Z(14),Z(15)/ '<=','>=',' ='/
```

```
C
C
C      READ AND PRINT PROBLEM NAME
C
10005 READ(5,10010)(PROB(I),I=1,20)
10010 FORMAT (20A4)
10015 WRITE(6,10020)(PROB(I),I=1,20)
10020 FORMAT(1H1,20A4,//)
C
C
C      READ PROBLEM ATTRIBUTES
C
C      ITYPE IS 1=MAX; 2=MIN
C      NREAL IS NO. OF REAL VARIABLES
C      NSLCK IS NO. OF SLACKS (.LE. CONSTRAINTS)
C      NSWA IS NO. OF ARTIFICIALS (.GE.CONSTRAINTS)
C      NEWA IS NO. OF EQUATION CONSTRAINTS (=)
C
 1000 READ(5,300) ITYPE, NREAL, NSLCK, NSWA, NEWA
  300 FORMAT(5I4)
C
C      END OF DATA TEST - CHECK FOR SECOND BLANK CARD
C
      IF (ITYPE.EQ.0) GO TO 600
      IF (ITYPE .GT. 2) GO TO 590
C
C
C      ECHO PROBLEM ATTRIBUTES
C
C
      NART = NSWA + NEWA
      Y9 = Z(1)
      IF (ITYPE .EQ. 2) Y9 = Z(2)
      WRITE (6,3391)
 3391 FORMAT (1H0,40X,' LP INPUT DATA',/42X,'*************',///)
      WRITE (6,10025) Y9,NREAL,NSLCK,NSWA,NSWA,NEWA,NART
10025 FORMAT(1H0,20X,'OBJECTIVE FUNCTION TO BE ',A3,'IMIZED WITH ',I4,'
     2DECISION VARIABLES',//20X,'SUBJECT TO :',/20X,I4,' <= CONSTRAINTS'
     3,1X,'(',I3,' SLACK VARIABLES )',/20X,I4,' >= CONSTRAINTS',1X,'(',I
     43,' SURPLUS VARIABLES )',/20X,I4,' = CONSTRAINTS',1X,'(',I3,' ART
     5IFICIAL VARIABLES )')
C
C      SET PARAMETERS
C
C      N1 IS NUMBER OF CONSTRAINTS
C      N2 IS TOTAL NUMBER OF VARIABLES
C      N4 IS LAST COLUMN NUMBER
C      N5 IS THE LAST ROW NUMBER
C      NART IS NO. OF ARTIFICIAL VARIABLES
C      NR3 IS COLUMN IN TABLE IN WHICH LAST REAL VARIABLE LIES
C      NRE1 IS FIRST COLUMN OF FICTICIOUS VARIABLES
C      KQ IS THE ROW NUMBER FOR OBJECTIVE FUNCTION COEFFICIENTS
C
      N1 = NSLCK + NSWA + NEWA
      N2 = NREAL + NSLCK + NEWA + 2 * NSWA
      N4=N2+3
      N5=N1+3
      NART=NSWA+NEWA
      NR3=NREAL+3
      NRE1=NR3+1
      KQ=N1+ITYPE
      N33=NR3+NSLCK+NSWA
C
      IF ( N4 .GT. 255 .OR. N5 .GT. 85) GO TO 510
C
C      INITIALIZE VARIABLES TO BE USED AS INDICES
C
      K1 = 0
      K3=0
      KJI=0
      N3=0
C
C
```

```
C      INITIALIZE VALUE OF ITERATION NUMBER
C
       ITNUM=0
C
C
C      DETERMINE WHICH ROW TO USE FOR OPPORTUNITY COST; WHETHER MAX OR MIN
C      IS DESIRED
C      KS IS ROW FOR OPPORTUNITY COST
C
       IF (ITYPE-1) 802,801,802
   801 KS=N1+2
       TYPE = -1.
       GO TO 8001
   802 KS=N1+1
       TYPE = 1.
  8001 CONTINUE
C
C
C*******************************************************************************
C
C
       WRITE (6,10030)
 10030 FORMAT (1H0,20X,'      CONSTRAINT COEFFICIENTS AND RIGHT HAND SIDE
      3'/ 20X,'*****************************************************'/)
C
C
C      BEGIN READING IN PROBLEM CONSTRAINTS
C
   803 K=0
       IFLAG=0
       DO 999 I = 1,N1
       READ(5,301) (A(I,J),J=4,NR3),A(I,3)
C
C
C      IFLAG IS ERROR FLAG SET IF NEGATIVE VALUE APPEARED ON RIGHT SIDE
C      OF A CONSTRAINT EQUATION.  PROGRAM WILL READ REMAINDER OF INPUT
C      DATA FOR THIS PROBLEM, PRINT ERROR MESSAGE, AND CONTINUE
       IF(A(I,3).LT.0.0) IFLAG=1
C
C
C      INITIALIZE VALUES IN COLUMN 2 OF TABLEAU AT 0
C
       A(I,2)=0
C
C
C      KK IS THE NUMBER OF THE CURRENT COLUMN
C
       K=K+1
       KK=NR3+K
C
C
C      TEST FOR END OF SLACKS.  IF NO MORE - GO AROUND
C
       IF (I - NSLCK) 60,60,69
C
C
C      PLACE VARIABLE NUMBER IN COLUMN  1
C
    60 A(I,1)=KK-3
C
C
C      SCAN THE ROW BEING WORKED ON AND INSERT A 1 IN THE COLUMN EQUAL TO
C      THE NUMBER OF VARIABLES + THE ROW NUMBER.  THIS WILL FORM A DIAGONAL
C      IN THE ARRAY.  THE REMAINING BLANKS ARE FILLED IN WITH ZEROES.
C
       DO 61 KKK=NRE1,N4
       IF (KK-KKK)63,62,63
    62 A(I,KKK)=1
       GO TO 61
    63 A(I,KKK)=0
```

```
   61 CONTINUE
      Y10 = Z(13)
      GO TO 4
C
C
C     READ IN ARTIFICIALS (.GE.)
C     IF NO ARTIFICIALS, GO AROUND
C
   69 IF(I - (NSLCK + NSWA)) 691,691,74
C
C
C     CONTINUE THE DIAGONAL  IN THE ARRAY WITH -1 IN THE ROWS CONTAINING
C     ARTIFICIALS.  PLACE A 1 IN THE COLUMN EQUAL TO THE NUMBER OF
C     CONSTRAINTS + THE NUMBER OF ARTIFICIALS + THE ROW NUMBER
C
  691 DO 70 KKK=NRE1,N4
      IF (KK-KKK)72,71,72
   71 A(I,KKK)=-1
      KJI=KKK+NSWA
      A(I,KJI)=1
C
C
C     PUT VARIABLE NUMBER IN FIRST COLUMN (ARTIFICIAL VARIABLE)
C
      A(I,1)=KJI-3
      GO TO 70
C
C
C     FILL IN ALL OTHERS WITH ZEROES
C
   72 IF (KJI-KKK)73,70,73
   73 A(I,KKK)=0
   70 CONTINUE
C
C
C     THESE CONSTRAINTS FINISHED READING IN
C
      Y10 = Z(14)
      GO TO 4
C
C
C     HAD THERE BEEN ANY ARTIFICIALS (.GE.) ?
C
   74 IF(NSWA) 744,744,745
C
C     IF THERE WERE, INCREMENT COLUMN INDEX
C
  745 KJI=KJI+1
      GO TO 746
C
C
C     READ IN EQUALITY CONSTRAINTS
C
C
C     PUT VARIABLE NUMBER IN FIRST COLUMN
C
  744 KJI=KK
  746 A(I,1)=KJI-3
C
C
C     FORM ANOTHER DIAGONAL BY PLACING A 1 IN THE COLUMN EQUAL TO THE
C     NUMBER OF VARIABLES + THE ROW NUMBER + THE NUMBER OF ARTIFICIALS
C
      DO 75 KKK=NRE1,N4
      IF (KJI-KKK) 77,76,77
```

```
C
   76 A(I,KKK)=1
      GO TO 75
   77 A(I,KKK)=0
   75 CONTINUE
C
      Y10 = Z(15)
C
C     END OF CONSTRAINT CO-EFF READ IN
C
C     ECHO CONSTRAINT COEFFICIENTS
C
C
    4 WRITE (6,10045) I
10045 FORMAT (1H0,20X,'CONSTRAINT NO.',I4)
      WRITE (6,10035)(A(I,J),J=4,NR3)
10035 FORMAT (1H ,20X,10F11.4)
      WRITE (6,10040) Y10,A(I,3)
10040 FORMAT (1H0,20X,A4,F11.4)
C
C
  999 CONTINUE
C
C
C***********************************************************************************
C
C
C     READ IN OBJECTIVE FUNCTION COEFFICIENTS
C
      READ(5,301) (B(J),J=4,NR3)
  301 FORMAT(10F8.0)
C
C
C     PROGRAM FINISHED READING INPUT DATA CARDS
C     IF IFLAG WAS SET TO 1 , GO TO NEXT DATA SET
      IF(IFLAG.EQ.1) GO TO 502
C
C
C     THE VECTOR 'B' STORES THE CO-EFFS OF THE OBJECTIVE FUNCTION.
C     IF IPH=1 THEN IT IS PHASE I,IF IPH=2 THEN IT IS PHASE II.
C
C
      IIPH=1
      IF (NSWA+NEWA) 51,51,52
   52 IPH=1
      KQ=N1+2
      KS=N1+1
      DO 58 J=4,NR3
   58 A(KQ,J)=0.
      GO TO 127
   51 KQ=N1+ITYPE
      IF (ITYPE-1) 756,756,757
  756 KS=N1+2
      GO TO 758
  757 KS=N1+1
  758 CONTINUE
      DO 54 J=4,NR3
   54 A(KQ,J)=B(J)
      IPH=2
      IF (IIPH.EQ.2) GO TO 55
C
C     ECHO OBJECTIVE COEFFICIENTS
C
  127 WRITE (6,10050)
10050 FORMAT (1H0,20X,'OBJECTIVE FUNCTION COEFFICIENTS',/21X,'*********
     5********************'/)
      WRITE (6,10055)(B(J),J=4,NR3)
10055 FORMAT (1H ,20X,10F11.4)
C
C
C     CALC NO. OF VARIABLES ADDED THAT WILL HAVE ZERO CO-EFFS IN OBJECTIVE
C
   55 K = NR3 + NSLCK + NSWA
C
C
```

```
C        FILL IN FICTICIOUS OBJECTIVE CO-EFFS WITH ZERO CO-EFFS
C
         DO 81 J=NRE1,K
      81 A(KQ,J)=0
C
C
C        ARE THERE ANY ARTIFICIAL VARIABLES?
C
         KKK = K + NSWA + NEWA
         IF (KKK-K)703,703,884
C
C
C
C        FORMULATE THE OBJECTIVE FUNCTIO FOR THE PHASE I.IT IS THE SUM OF
C        ARTIFICIAL VARIABLES.
C
C
C        IF IT IS PHASE II THEN FORMULATE THE (MODIFIED) OBJECTIVE FUNCTION
C        ( A LARGE CO-EFF IS ATTACHED IN THE OBJECTIVE FUNCTION TO THOSE
C        VARIABLES WHICHUHAVE NEGATIVE NET EVALUATION VALUES IN PHASE I)
C
C
     884 KK=K+1
         DO 882 J=KK,KKK
         IF (IPH.EQ.1) GO TO 56
         A(KQ,J)=B(J)
         GO TO 882
      56 A(KQ,J)=1.
     882 CONTINUE
C
C
C        FIND COLUMN REPRESENTING VARIABLE
C
     703 DO 701 I=1,N1
         KR =A(I,1)+3.
C
C
C        INSERT OBJECTIVE CO-EFFS FOR PROGRAM VARIABLES
C         IN PROPER ROW DEPENDING ON MAX OR MIN
C
     701 A(I,2)=A(KQ,KR)
     700 A(N1 + 1,3) = 0.
C
C
C        CALCULATE TOTAL COST
C
         DO 150 I=1,N1
     150 A(N1 + 1,3) = A(N1 + 1,3) + A(I,2)*A(I,3)
C
C        THIS DO-LOOP (OUTER) BUILDS NET EVALUATION VALUES
C
         DO 154 J=4,N4
         A(KS,J)=0
C
C        THIS DO-LOOP (INNER) BUILDS GROSS EVALUATION VALUES
C
         DO 155 I=1,N1
     155 A(KS,J)=A(KS,J)+A(I,2)*A(I,J)
C
C
     154 A(N5,J) =  A(N1 + 1,J) - A(N1 + 2,J)
C
C
C        INCREMENT ITERATION NUMBER
C
      23 ITNUM=ITNUM+1
C
C
C*****************************************************************************
C
C
```

```
C        DETERMINE IF OPTIMAL SOLUTION
C        CHECK NET EVALUATION ROW FOR ALL ZEROES OR NEGS
C        IF ALL ZEROES OR LESS, WRITE OUT SOLUTION
C        IF NOT, BRANCH TO SEARCH ROUTINE
C
 2333 DO 1 IV=4,N4
      IF(A(N5,IV).GT. .0 ) GO TO 2
    1 CONTINUE
C
C
C        CHECK FOR INFEASIBILITY
C
C
C        THE VARIABLES WHICH HAVE NEGATIVE NET EVALUATION VALUES AT THE END
C        OF PHASE I SHOULD BE KEPT OUT OF PHASE II.
C
C        THIS IS ACHIEVED BY ATTACHING A LARGE COEFFICIENT TO THESE
C        VARIABLES IN PHASE II OBJECTIVE FUNCTION.
C
C
      IF (IPH.EQ.2) GO TO 91
      YY9=0.
      NONAVR=NREAL+NSLCK+NSWA
      DO 755 KI=1,N1
      IF (YY9.GT.A(KI,1)) GO TO 755
      YY9=A(KI,1)
      NUMI=YY9
      IF(NUMI.LE.NONAVR) GO TO 755
      IF(A(KI,3).GT. .0) GO TO 552
  755 CONTINUE
      IWARN=0.
      DO 7761 KM=4,N4
      IF(ITYPE-1) 7762,7762,7763
 7762 IF(A(N5,KM)) 7765,7764,7764
 7764 IF (KM.LE.N33) GO TO 7761
      B(KM)=0.
      GO TO 7761
 7765 B(KM)=-9.9E+48
      GO TO 7761
 7763 IF (A(N5,KM)) 7767,7766,7766
 7766 IF (KM.LE.N33) GO TO 7761
      B(KM)=0.
      GO TO 7761
 7767 B(KM)= 9.9E+48
 7761 CONTINUE
      IPH=2
      IIPH=2
      GO TO 51
C
C*************************************************************************************
C
   91 IWARN = 0
      WRITE(6,10020)(PROB(I),I=1,20)
      WRITE (6,3392)
 3392 FORMAT(1H0,41X,'LP SOLUTION',/42X,'***********',////)
C
C        PRINT VALUE OF OBJECT FUNCTION
C
      Y9 = Z(1)
      IF(ITYPE .EQ. 2) Y9 = Z(2)
12005 FORMAT(1H0,20X,'OBJECTIVE FUNCTION ',A3,'IMIZED AT ',1PE14.5)
12010 WRITE (6,12005) Y9,A(N1 + 1,3)
C
C
C        PRINT NUMBER OF FINAL ITERATION
C
13010 WRITE(6,13005) ITNUM
13005 FORMAT(1H0,20X,I3,' ITERATIONS REQUIRED')
C
C
C        SEQUENCE AND PRINT REAL AND SLACK OR SURPLUS VARIABLES (NUMBER AND
C        QUANTITY)
C
C        NONAVR = NO. OF NON-ARTIFICIAL VARIABLES
C
      NONAVR = NREAL + NSLCK + NSWA
```

```
14005 FORMAT(1H0,20X,'INCLUDED VARIABLES'/1H ,20X,'*******************')
19005 WRITE(6,14005)
14015 FORMAT(1H0,17X,'VARIABLE',4X,'QUANTITY OF',4X,'VARIABLE',6X,'ASSOC
     5IATED WITH'/1H ,18X,'NUMBER',4X,'THIS VARIABLE',5X,'TYPE',7X,'CONS
     5TRAINT NUMBER')
      WRITE(6,14015)
      DO 14090 NLOOP=1,N1
      Y9=9.9E+48
      DO 14045 I=1,N1
      IF(Y9.LT.A(I,1)) GO TO 14045
      Y9=A(I,1)
      NUMZ = I
14045 CONTINUE
      NUM=Y9
      MNUM = NUM - NREAL
      IF(NUM .LE. NREAL) GO TO 14091
      IF (NUM .GT. NONAVR) GO TO 550
      IF(NUM .GT. NREAL + NSLCK) GO TO 14092
      WRITE(6,14060) NUM,A(NUMZ,3),Z(3),Z(4) ,Z(12),MNUM
      GO TO 14085
14091 WRITE(6,14060)NUM,A(NUMZ,3), Z(7),Z(8)
      GO TO 14085
  550 IWARN = 1
      IF (ABS(A(NUMZ,3)).GT. .00005 ) IWARN=2
      MNUM = MNUM-NSWA
      WRITE (6,14060) NUM, A(NUMZ,3), Z(9), Z(10), Z(11),MNUM
      GO TO 14085
14092 WRITE(6,14060) NUM, A(NUMZ,3), Z(5), Z(6),Z(12),MNUM
14060 FORMAT(1X,/,20X,I3,6X,1PE10.4,6X,3A4,2X,I3)
14085 A(NUMZ,1) = 9.9E+50
14090 CONTINUE
      IF (IWARN .EQ. 0) GO TO 5001
      IF (IWARN .EQ. 2) GO TO 552
 5001 CONTINUE
      WRITE(6,3393)
 3393 FORMAT(1X,//////,21X,'ASSOCIATED DUAL VARIABLES',/21X,'************
     C************* ',/)
      WRITE(6,7723)
 7723 FORMAT(18X,'CONSTRAINT',11X,'VALUE OF THE ')
      WRITE (6,7724)
 7724 FORMAT(22X,'NO.',14X,'DUAL VARIABLE',/)
      ND=NR3+NSLCK
C
C
C     WRITE OUT THE VALUES OF DUAL VARIABLES FROM THE ROW THAT STORES
C     THE GROSS EVALUATION VALUES.
C
C
      IKS=1
      IF (NSLCK.EQ.0)  GO TO 7734
      DO 7740 KD=NRE1,ND
      IF (ITYPE-1) 7774,7774,7775
 7774 BGG=A(KS,KD)
      GO TO 7776
 7775 BGG=-A(KS,KD)
 7776 CONTINUE
      WRITE(6,7742) IKS,BGG
 7742 FORMAT(20X,I3,15X,F10.5)
      IKS=IKS+1
 7740 CONTINUE
 7734 CONTINUE
      IF (NART.EQ.0) GO TO 7705
      NDD=NRE1+NSLCK+NSWA
      DO 7735 KD=NDD,N4
      IF (ITYPE-1) 7874,7874,7875
 7874 BGG=A(KS,KD)
      GO TO 7876
 7875 BGG=-A(KS,KD)
```

```
   7876 CONTINUE
        WRITE(6,7742) IKS,BGG
        IKS=IKS+1
   7735 CONTINUE
   7705 CONTINUE
      IF (IWARN .EQ. 0) GO TO 500
      WRITE (6,553)
    553 FORMAT (//,20X,' **** NOTE ****'//,20X,' THIS PROBLEM MAY CONTAIN
       1REDUNDANT CONSTRAINTS')
      GO TO 500
    552 WRITE(6,10020)(PROB(I),I=1,20)
      WRITE (6,3392)
      WRITE (6,551)
    551 FORMAT(1X,////////////////40X,'*******************************',/40X,
       C'*',28X,'*',/40X,'*',28X,'*',/40X,'*     NO FEASIBLE SOLUTION      *'
       C,/40X,'*',28X,'*',/40X,'*',28X,'*',/40X,'*******************************
       C*****')
C
C
C       GO TO NEXT PROBLEM
C
        GO TO 500
C
C
C*************************************************************************************
C
C
C       BEGINNING OF THE PIVOT ROUTINE
C
C       WAS BRANCH HERE CAUSED BY LAST ITEM?
C       IF NO, CONTINUE SEARCH
C       IF YES, USE THE LAST COLUMN AS PIVOT COLUMN
C
      2 IF (IV-N4) 3,44,3
C
C
C       CONTINUE SEARCH TO END OF REALS FOR GREATEST VALUE
C       SAVE POINTER TO COLUMN OF GREATEST REAL VARIABLE
C
      3 J=IV+1
        DO 5 K=J,N4
        IF(A(N5,K))5,5,6
      6 IF(A(N5,IV)- A(N5,K))7,5,5
      7 IV=K
      5 CONTINUE
C
C
C       DETERMINE PIVOT VALUE - SEARCH DOWN CHOSEN COLUMN
C
     44 DO 8 LV=1,N1
        IF(A(LV,IV))8,8,9
      8 CONTINUE
C
C
C       NO POSITIVE VALUE IN COLUMN IMPLIES UNBOUNDED SOLUTION
C       TERMINATE THIS PROBLEM IF UNBOUNDED SOLUTION
C
      WRITE(6,10020)(PROB(I),I=1,20)
      WRITE (6,3392)
      WRITE (6,303)
    303 FORMAT(1X,////////////////40X,'*******************************',/40X,
       C'*',28X,'*',/40X,'*',28X,'*',/40X,'*     UNBOUNDED SOLUTION        *'
       C,/40X,'*',28X,'*',/40X,'*',28X,'*',/40X,'*******************************
       C*****')
C
        GO TO 500
C
C
C       DID SEARCH FOR + VALUE END IN LAST ROW OF SELECTED COLUMN?
C       IF YES (IT IS THE ONLY +), DON'T BOTHER TO SEARCH FOR HIGHEST +
C
      9 IF(LV-N1)106,107,106
C
C
```

```
C      BEGIN SEARCH FOR HIGHEST POSITIVE RATIO TO DETERMINE PIVOT VALUE
C
  106 R1=A(LV,3)/A(LV,IV)
      J=LV+1
      DO 10 K=J,N1
      K11=LV
      IF (A(K,IV))10,10,11
   11 R2=A(K,3)/A(K,IV)
      IF(R1-R2) 10,200,12
  200 DO 21 KJ=4,N33
      K22=K
      B1B=A(K11,KJ)/A(K11,IV)
      B2B=A(K22,KJ)/A(K22,IV)
      IF (B1B-B2B) 24,25,26
   24 LV=K11
      GO TO 10
   25 GO TO 21
   26 LV=K22
      GO TO 10
   21 CONTINUE
   12 R1=R2
      LV=K
   10 CONTINUE
C
C
C      TRANSFORM NON-KEY ROWS
C
C      OUTER DO-LOOP RUNS DOWN ALL ROWS EXCEPT PIVOT ROW
C      Q IS RATIO OF (CURRENT ROW)/(PIVOT ROW) FOR PIVOT COLUMN
C
  107 DO 13 I=1,N1
      IF(I-LV)14,13,14
   14 Q=A(I,IV)/A(LV,IV)
C
C      INNER DO-LOOP RUNS ACROSS EACH ROW
C      A(LV,J) IS CORRESPONDING PIVOT ROW MEMBER FOR CURRENT COLUMN
C
      DO 15 J=3,N4
   15 A(I,J)=A(I,J)-(A(LV,J)*Q)
C
   13 CONTINUE
C
C
C      TRANSFORM PIVOT ROW
C      INSERT NEW VARIABLE NUMBER & OBJECTIVE CO-EFF
C
      A(LV,1)=IV-3
      A(LV,2)=A(KQ,IV)
      QQ=A(LV,IV)
C
C
C      DIVIDE ROW THROUGH BY PIVOT VALUE
C
      DO 16 J=3,N4
   16 A(LV,J)=A(LV,J)/QQ
C
C
C      CALCULATE NEW TOTAL COST VALUE
C
      A(N1 + 1,3) = 0.
      DO 960 I=1,N1
  960 A(N1+ 1,3) = A(N1 + 1,3) + A(I,2)*A(I,3)
C
C
```

```
C       TRANSFORM GROSS AND NET EVALUATION ROWS
C
        DO 20 J=4,N4
        A(KS,J)=0
        DO 19 I=1,N1
   19 A(KS,J)=A(KS,J)+A(I,2)*A(I,J)
   20 A(N5,J) = A(N1 + 1,J) - A(N1 + 2,J)
        GO TO 23
C
C
C       RETURN TO EVALUATE THIS ITERATION
C
C
C       RETURN TO READ NEXT SET OF DATA (THE NEXT PROBLEM)
C
  500 GO TO 10005
C
  502  WRITE(6,503)
  503  FORMAT(1H0,'*** WARNING ***',
     1/' YOU HAVE INPUT A NEGATIVE QUANTITY ON THE RIGHT SIDE OF ',
     2'A CONSTRAINT EQUATION.',
     3 /' PROGRAM PROCEEDS TO NEXT DATA SET.')
        GO TO 10005
C
  510 WRITE (6,511) N4,N5
  511 FORMAT (1H0,20X,'***** WARNING *****',//20X,'THIS PROBLEM REQUIRES
     1',I4,' COLUMNS AND',I4,' ROWS.'/1H0,20X,'THIS EXCEEDS MATRIX SIZE
     2OF 254 COLS. AND 84 ROWS.'/1H0,20X,'THE PROGRAM WILL ATTEMPT TO FL
     3USH THIS PROBLEM AND CONTINUE.')
C
        NFL = N1 * (N2 + 1) + N2
        NCARD = NFL/10
        IF (MOD(NFL,10) .NE. 0) NCARD = NCARD + 1
        DO 512 JFL = 1,NCARD
        READ (5,513) X
  513 FORMAT (A4)
  512 CONTINUE
        GO TO 10005
C
  590 WRITE (6,595)
  595 FORMAT (1H0,20X,'*****TYPE CANNOT EXCEED 2*****CHECK PROBLEM ATTRI
     5BUTES*****'/1H ,30X,'**********EXECUTION TERMINATED**********')
C
  600 STOP
        END
```

BIBLIOGRAPHY

The bibliography presented here contains a selection of books that should provide sufficient material for the serious scholar to design a course of self-study. Each book listed is accompanied by an annotation indicating the level of mathematical sophistication required and any special features that make the book of particular interest.

Group I

Books requiring a minimal mathematical background.

1. Bierman, Harold Jr., Charles P. Bonini and Warren H. Hausman, *Quantitative Analysis for Business Decisions*, 4th ed., Homewood, Ill.: Richard D. Irwin, 1973.

    This book contains a survey of quantitative techniques used in business and industry and devotes only three chapters to linear programming. An approach which is based on linear algebra is treated in an appendix to one of the chapters. This book is heavily oriented toward problem formulation and includes numerous problems.

2. Daellenbach, Hans G., and Earl J. Bell, *User's Guide to Linear Programming*, Englewood Cliffs, N. J.: Prentice-Hall, 1970.

    This basic text provides an excellent conceptual approach to all major topics in linear programming. The chapter on problem formulation is a strong point of the book. In addition, there is a section on computer solution of linear programming problems, including a listing of a linear programming code written in FORTRAN IV.

3. Driebeek, Norman J., *Applied Linear Programming*, Reading, Mass.: Addison-Wesley Publishing Company, 1969.

    This text is problem description oriented. Detailed examples and case studies are included to illustrate the role of the computer in solving linear programming problems.

4. Frazer, Ronald J., *Applied Linear Programming*, Englewood Cliffs, N. J.: Prentice-Hall, 1968.

    Special features of interest in this book are the FORTRAN II program developed by the author and a chapter devoted to model formulation.

5. Gass, Saul J., *An Illustrated Guide to Linear Programming*, New York: McGraw-Hill Book Co., 1970.

    This book presents the fundamentals of linear programming in a simple and entertaining fashion. The only solution techniques discussed in the text are graphical methods with the Simplex method being treated briefly in an appendix. This is a book for the true novice who has very little mathematical background. An extensive bibliography of linear programming applications is provided.

6. Hughes, Ann J., and Dennis E. Grawoig, *Linear Programming: An Emphasis on Decision Making*, Reading, Mass.: Addison-Wesley Publishing Co., 1973.

    An excellent conceptual introduction to linear programming applications and the Simplex method. Special features of the text include chapters on goal programming and integer programming. There are many problems at the end of each chapter, and one chapter devoted to case study situations.

7. *Introduction to Linear Programming*, White Plains, New York: IBM Technical Publication, 1964.

    This is an introductory manual that illustrates a variety of problems that can be solved using linear programming. A large-scale problem is presented, and examples of typical computer output are included. The manual contains a glossary that defines the technical terms associated with linear programming.

8. Levin, Richard I., and Rudolph P. Lamone, *Linear Programming for Management Decisions*, Homewood, Ill.: Richard D. Irwin, 1969.

    This book provides a rather extensive treatment of linear programming at the elementary level. A feature of particular interest to the practitioner is a good chapter on management applications.

9. Levin, Richard I., and Charles A. Kirkpatrick, *Quantitative Approaches to Management*, New York: McGraw-Hill Book Co., 1971.

This is a good text for the nonquantitatively oriented reader who would like some familiarity with the quantitative tools available to management. Three chapters are devoted to linear programming and the special cases of the transportation and assignment problems. Excellent use is made of example problems in the development of the theory.

10. Loomba, N. Paul, *Linear Programming*, New York: McGraw-Hill Book Co., 1964.

A unique feature of this introductory linear programming book is the author's discussion of "systematic trial and error methods".

11. Martin, E. W. Jr., *Mathematics for Decision Making*, Homewood, Ill.: Richard D. Irwin, 1969.

The treatment of linear programming provided by this text is the fifth part of a programmed instruction text on basic mathematics. All of the necessary background in algebra and linear equations is contained in the earlier sections of the book. If you like programmed instruction, you will find this book is especially well done.

12. Meisels, Kurt, *A Primer of Linear Programming*, New York: New York University Press, 1962.

Written for business managers and executives, this book presents graphical methods for two variable problems and an introduction to the Simplex method.

13. Metzger, Robert W., *Elementary Mathematical Programming*, New York: John Wiley & Sons, 1958.

The emphasis of this paperback is on the managerial implications of linear programming. It includes sections on production planning, job evaluation, and material handling.

14. Naylor, Thomas H., Eugene T. Byrne, and John M. Vernon, *Introduction to Linear Programming*, Belmont, Cal.: Wadsworth Publishing Co., 1971.

>   The necessary background in linear equations and Gauss-Jordan elimination is developed as needed. In addition to the usual chapter on the transportation problem, additional features of this book are chapters on integer programming, input-output analysis, and game theory.

15. Richmond, Samual B., *Operations Research for Management Decisions*, New York: The Ronald Press Co., 1968.

>   Four chapters of this introductory operations research text are devoted to linear programming.

16. Stockton, R. S., *Introduction to Linear Programming*, Homewood, Ill.: Richard D. Irwin, 1971.

>   A short paperback, this book devotes a chapter to developing the necessary background in linear equations. Its strong points are its managerial orientation and the self-checking exercises at the end of the sections.

17. Thompson, Gerald E., *Linear Programming*, New York: The MacMillan Co., 1971.

>   This book contains an extensive treatment of linear programming at a modest mathematical level. The author deals almost exclusively with graphical methods of solution. Of particular interest are the special topics covered: concave separable programming, multiple objectives, integer programming, and linear programming with uncertainty.

Group II

Books for the more mathematically inclined.

1. Carr, Charles R., and Charles W. Howe, *Quantitative Decision Procedures in Management and Economics*, New York: McGraw-Hill Book Co., 1964.

>   A text in deterministic models, this book devotes about 100 pages to the treatment of linear programming. In a rather concise style, this text includes a presentation of the Simplex method and duality theory.

2. Charnes, A., and W. W. Cooper, *Management Models and Industrial Applications of Linear Programming*, Volumes I and II, New York: John Wiley & Sons, 1961.

    This book provides an extensive investigation of linear programming methods at a moderate mathematical level. A chapter is devoted to developing the concepts from matrix or linear algebra used in the sequel. Of particular interest are the discussions of goal programming and the chapter devoted to the resolution of degeneracy.

3. Charnes, A., W. W. Cooper, and A. Henderson, *An Introduction to Linear Programming*, New York: John Wiley & Sons, 1953.

    This paperback is one of the earliest publications on linear programming. It is divided into two parts. The first part, "An Economic Introduction to Linear Programming", requires only a moderate mathematical background. However, the second part, "Lectures on the Mathematical Theory of Linear Programming", requires a good deal more mathematical maturity.

4. Churchman, C. W., R. L. Ackoff, and E. L. Arnoff, *Introduction to Operations Research*, New York: John Wiley & Sons, 1957.

    This book, which is one of the classics in operations research, devotes approximately 100 pages to linear programming. It develops and illustrates many of the important applications of operations research with a heavy emphasis on model formulation.

5. Dantzig, G. B., *Linear Programming and Extensions*, Princeton, N. J.: Princeton University Press, 1963.

    Many consider this book to be "the" linear programming book. Although much of the material is difficult reading, primarily because of the older linear algebra approach, the wealth of material makes it a must for anyone that wants an in-depth understanding of linear programming. The problems at the end of each chapter are among the best to be found. An extensive bibliography of work prior to 1963 is included.

6. Gale, David, *The Theory of Linear-Economic Models*, New York: McGraw-Hill Book Co., 1960.

> An extensive treatment of linear programming with extensions to game theory is provided by this text. The treatment is at a rigorous mathematical level, hence a good deal of mathematical maturity should be a prerequisite to study of this book.

7. Garvin, Walter W., *Introduction to Linear Programming*, New York: McGraw-Hill Book Co., 1960.

> This text does an especially good job on sensitivity analysis and parametric linear programming. No matrix notation is used, but where necessary the author presents detailed rigorous proofs of theorems.

8. Gass, S., *Linear Programming*, 3rd ed., New York: McGraw-Hill Book Co., 1969.

> A feature of particular interest in this very thorough linear programming text is the chapter the author devotes to parametric linear programming.

9. Gue, Ronald L., and Michael E. Thomas, *Mathematical Methods in Operations Research*, New York: The MacMillan Co., 1968.

> This text provides a survey of many important operations research techniques. The treatment of linear programming is concise (39 pages); a theorem-proof approach is used. Linear programming through duality theory is treated, and several problems, some of a theoretical nature, are given.

10. Hadley, G., *Linear Programming*, Reading, Mass.: Addison-Wesley Publishing Co., 1962.

> This book provides one of the most extensive treatments available on linear programming. Strong points are its in-depth coverage and its numerous theoretically oriented problems.

11. Hillier, F., and G. J. Lieberman, *Introduction to Operations Research*, San Francisco: Holden-Day, 1967.

This book provides a survey of operations research techniques with three chapters devoted to linear programming. The first two chapters require less mathematical sophistication and take the reader through an introduction to duality theory. The last chapter treats advanced theory such as parametric linear programming and is presented at a higher level of mathematical sophistication.

12. Hu, T. C., *Integer Programming and Network Flows*, Reading, Mass.: Addison-Wesley Puglishing Co., 1969.

Approximately the first 100 pages of this high-level text are devoted to linear programming. The sections on complementary slackness and duality are excellent. The remainder of the text explores network flow concepts and integer programming.

13. Karlin, Samuel, *Mathematical Models and Theory in Games, Programming and Economics*, Vol. I, Reading, Mass.: Addison-Wesley Publishing Co., 1959.

A rigorous treatment of linear programming is contained in this book devoted to mathematical methods in economics. It is not recommended for the mathematically unsophisticated scholar. A mathematical treatment of game theory and its connection with linear programming is a strong point of the book.

14. Kim, Chaiho, *Introduction to Linear Programming*, New York: Holt, Rinehart and Winston, 1971.

The geometrical and computational aspects of linear programming are covered in depth. The text includes many problems for student practice and an extensive bibliography, including about 100 articles on applications.

15. Llewellyn, Robert W., *Linear Programming*, New York: Holt, Rinehart and Winston, 1964.

This book provides an extensive treatment of linear programming theory. The revised simplex and primal-dual methods are covered. In addition a chapter showing the application of linear programming to game theory is included.

16. Luenberger, David G., *Introduction to Linear And Nonlinear Programming*, Reading, Mass.: Addison-Wesley Publishing Co., 1973.

> A comprehensive mathematical treatment of linear programming is provided in Part 1 of this text. The text is oriented toward preparing the reader for advanced theoretical studies in optimization methods.

17. Orchard-Hays, William, *Advanced Linear Programming Computing Techniques*, New York: McGraw-Hill Book Co., 1968.

> Some previous knowledge of linear programming as well as a familiarity with modern electronic computing techniques is assumed. A strong point of the book is its treatment of the computing difficulties associated with large-scale linear programming problems.

18. Simmons, Donald M., *Linear Programming For Operations Research*, San Francisco: Holden-Day, Inc., 1972.

> This text provides a comprehensive treatment of linear programming using a theorem building approach. An undergraduate course in differential calculus and a first course in linear algebra are the assumed prerequisites. Exercises in the text are mainly intended to strengthen the student's knowledge of the theorems presented.

19. Smyth, William R. Jr., and Lynwood A. Johnson, *Introduction to Linear Programming with Applications,* Englewood Cliffs, N. J.: Prentice-Hall, 1966.

> Based on a theorem-proof approach, this book gives an introductory treatment of linear programming through the development of the Simplex algorithm. A separate chapter on linear algebra provides the necessary background in this area.

20. Spivey, W. Allen, and Robert M. Thrall, *Linear Optimization,* New York: Holt, Rinehart and Winston, 1970.

> A fresh outlook in terms of a matrix approach to theorem proving is a strength of this text. Numerous problems are included at the end of each chapter.

21.   Strum, Jay E., *Introduction to Linear Programming*, San Francisco, Cal.: Holden-Day, Inc., 1972.

     Although this text is theory oriented, it does not take a formal theorem proving approach. Unique features of the book include an attempt to motivate duality by first developing sensitivity analysis, and an introductory chapter on graphs and networks. Very few exercises are provided at the end of each chapter.

22.   Teichroew, Daniel, *An Introduction to Management Science*, New York: John Wiley & Sons, 1966.

     This book, devoted to the application of the methods of management science in business, contains three chapters dealing with linear programming. The treatment of linear programming extends through duality theory.

23.   Wagner, H., *Principles of Operations Research with Applications to Managerial Decisions*, Englewood Cliffs, N. J.: Prentice-Hall, 1969.

     The author presents an introduction to nearly all of the operations research techniques in use today. Even so, the four-chapter treatment of linear programming is extensive.

*

INDEX

End of Volume